第二版

投考公務員 英文運用 模擬試卷精讀

Use of english: mock paper

Fong Sir 著

【目錄】

PART ONE 英文運用模擬試卷

PART TWO 模擬試卷答案

PART ONE
英文運用試模擬卷

CC-CRE-UE

文化會社出版社 **CULTURE CROSS LIMITED**

答題紙 ANSWER SHEET

請在此貼上電腦條碼 Please stick the barcode label here

(1) 考生編號 Candidate No.

(2) 考生姓名 Name of Candidate

(3) 考生簽署 Signature of Candidate

宜用H.B.鉛筆作答
You are advised to use H.B. Pencils

考生須依照下圖所示填畫答案：

23 A B C D E

錯填答案可使用潔淨膠擦將筆痕徹底擦去。

切勿摺皺此答題紙

Mark your answer as follows:

23 A B C D E

Wrong marks should be completely erased with a clean rubber.

DO NOT FOLD THIS SHEET

1	A	B	C	D	E	21	A	B	C	D	E
2	A	B	C	D	E	22	A	B	C	D	E
3	A	B	C	D	E	23	A	B	C	D	E
4	A	B	C	D	E	24	A	B	C	D	E
5	A	B	C	D	E	25	A	B	C	D	E
6	A	B	C	D	E	26	A	B	C	D	E
7	A	B	C	D	E	27	A	B	C	D	E
8	A	B	C	D	E	28	A	B	C	D	E
9	A	B	C	D	E	29	A	B	C	D	E
10	A	B	C	D	E	30	A	B	C	D	E
11	A	B	C	D	E	31	A	B	C	D	E
12	A	B	C	D	E	32	A	B	C	D	E
13	A	B	C	D	E	33	A	B	C	D	E
14	A	B	C	D	E	34	A	B	C	D	E
15	A	B	C	D	E	35	A	B	C	D	E
16	A	B	C	D	E	36	A	B	C	D	E
17	A	B	C	D	E	37	A	B	C	D	E
18	A	B	C	D	E	38	A	B	C	D	E
19	A	B	C	D	E	39	A	B	C	D	E
20	A	B	C	D	E	40	A	B	C	D	E

文化會社出版社
投考公務員 模擬試題王

英文運用
模擬試卷（一）

時間：四十分鐘

考生須知：

（一）細讀答題紙上的指示。宣布開考後，考生須首先於適當位置貼上電腦條碼及填上各項所需資料。宣布停筆後，考生不會獲得額外時間貼上電腦條碼。

（二）試場主任宣布開卷後，考生請檢查試題冊及確定試題冊內的試題。最後會有「**全卷完**」的字眼。

（三）本試卷各題佔分相等。

（四）**本試卷全部試題均須回答**。為便於修正答案，考生宜用HB鉛筆把答案填畫在答題紙上。錯誤答案可用潔淨膠擦將筆痕徹底擦去。考生須清楚填畫答案，否則會因答案未能被辨認而失分。

（五）每題只可填畫**一個**答案。如填劃超過一個答案，該題將**不獲評分**。

（六）答案錯誤，不另扣分。

（七）未經許可，請勿打開試題冊。

CC-CRE-UE

I. Comprehension (10 questions)

Instruction:

This section aims to test candidates' ability to comprehend a written text. A prose passage of non-technical background is cited. Candidates are required to exercise skills in deciding on the gist, identifying main points, drawing inferences, distinguishing facts from opinion, interpreting figurative language, etc.

Passage:

Jacob hated finishing things almost as much as he loved starting them. As a result, he had gotten into a million hobbies and activities, but he never stuck with any of them long enough to get any good.

He begged his mother for months for a guitar so that he could play Black Eyed Peas songs to Angie, a girl he liked, but after he finally got one for Christmas, he found out that guitars don't play themselves. He took a few lessons, but strumming the strings hurt his fingers and he didn't like holding the pick, so now the five-hundred dollar guitar lives under his bed.

After reading an ad in the back of one of his comic books, Jacob decided that he wanted a Wonder-Sweeper 5,000 metal detector, so that he could find buried pirate treasure. So he mowed lawns all summer and didn't spend his money on ice-cream like his younger brother, Alex. He saved it all in a shoe box in his closet. Then he shoveled driveways all winter, and he didn't spend his money on candy and chips like his classmates. By the time spring came he had saved $200, and he purchased the Wonder-Sweeper 5000 metal detector. He beeped it around the park for a while, be he soon found out that no pirates had ever set sail in his neighborhood, and if they had they didn't leave any treasure. Even though he found a key ring, forty-seven cents, and all the bottle caps he could throw, he buried the metal detector in his closest.

Given Jacob's history with hobbies, it was no surprise that Jacob's father was reluctant to buy him a magician's kit for his birthday. "Geez, Jacob... You sure you wouldn't rather I got you more guitar lessons?" He suggested. Jacob was insistent. "Dad, you've got to get me the magician's kit. This time I'll stick with it for real. I promise! Come on, Dad," Jacob begged. Jacob's father sighed and then replied, "Oh, I don't know, Jacob. Things are awfully tight right now." But Jacob's father was reminded of his own youth long ago, when he quit football and started karate practice before hardly getting his equipment dirty. So when Jacob's birthday came around, Jacob was both surprised and pleased to find the magician's kit that he had desired so badly with a big bright bow on it.

Jacob opened up the box and unwrapped the many parts in the kit. As he did so, he imagined sawing his pet cat in half and putting it back together to the amazement of his friends and family. He took the many fake coins, trick cards, and rope pieces of varying length on the kitchen table and imagined pulling rabbits out of his hat and turning them into pigeons with a mysterious puff of smoke. As Jacob continued pulling plastic thumbs, foam balls, and giant playing cards out of the magic kit, a commercial on the TV caught his attention. "Hey kids! Have you ever wanted to go to space? Experience what it's like to be an astronaut? Do you want to explore the universe? Well, now you can." As the commercial continued playing, Jacob walked away from the magic kit on the kitchen table and stared at the TV screen longingly. "For only $195 you can go to space camp and live life like an astronaut for a whole weekend. Enroll now for a once in a life time experience." Jacob's cry rang throughout the house as he yelled, "MOM!" He now knew what his true purpose in life was.

Questions:

1. **According to the text, why does Jacob stop playing the guitar?**

    A. It hurts his fingers.

    B. He'd rather play drums.

    C. It was too easy.

    D. He failed math.

2. **To whom did Jacob want to play Black Eyed Peas songs?**

    A. Alex

    B. Angie

    C. Mom

    D. Dad

3. **According to the passage, why does Jacob decide that he wants a metal detector?**

   A. He sees a man at the park with one.

   B. His father had one as a child.

   C. He saw a TV commercial for one.

   D. He read an ad for one in a comic book.

4. **How does Jacob get the items that he wants in the story?**

   A. He asks his mom.

   B. He asks his dad.

   C. He shovels driveways and mows lawns.

   D. He does all of these things to get what he wants.

5. **When did Jacob buy the metal detector?**

   A. In the fall

   B. In the summer

   C. In the spring

   D. In the winter

6. **True or False: The metal detector was a good investment for Jacob.**

   A. True

   B. False

7. **Why doesn't Jacob's father want to get him the magician's kit for his birthday?**

   A. Jacob failed math class.

   B. Jacob quits too many expensive activities.

   C. Jacob has been mean to his younger brother.

   D. Jacob went to the park without permission.

8. **Why does Jacob's father buy Jacob the magician's kit?**

   A. Jacob mowed the lawn.

   B. Jacob reminded his father of himself.

   C. Jacob bought ice cream for his brother.

   D. Jacob found his father's key ring.

9. **Which word is closest in meaning to the italicized word in the following sentence from paragraph four: "It was no surprise that Jacob's father was reluctant to buy him a magician's kit for his birthday"?**

   A. Happy

   B. Willing

   C. Proud

   D. Hesitant

10. What distracts Jacob from the magician's kit?

A. A TV commercial

B. His father

C. The kitchen table

D. A comic book

II. Error Identification (10 questions)

Instruction:

Knowledge on use of the language is tested through identification of language errors which may be lexical, grammatical or stylistic.

Questions:

11. **He was quite amusing when he heard what had happened.**

    A. was

    B. amusing

    C. heard

    D. had

    E. No error

12. **Turn left by the crossroads when you reach it.**

    A. Turn

    B. by

    C. when

    D. it

    E. No error

13. **He has been working here for sometimes.**

A. been

B. here

C. for

D. sometimes

E. No error

14. **He stopped to see if he could picked up the trail.**

A. stopped

B. see

C. could

D. picked

E. No error

15. **Although he jumped aside, but the stone hit him.**

A. jumped

B. but

C. the

D. hit

E. No error

16. I decided to climbed to the top of the hill to get a better view.

A. decided

B. climbed

C. get

D. view

E. No error

17. He jumped down after shouted a warning to those standing below.

A. jumped

B. shouted

C. to

D. standing

E. No error

18. After a few minutes, I look up and saw that it was getting dark.

A. few

B. look

C. saw

D. was

E. No error

19. I saw the blind man crossed the busy road without any help.

A. saw

B. crossed

C. without

D. any

E. No error

20. The robber gave the victim with a hard blow.

A. gave

B. with

C. a

D. blow

E. No error

III. Sentence Completion (10 questions)

Instruction:

In this section, candidates are required to fill in the blanks with the best options given. The questions focus on grammatical use.

Questions:

21. **Today Wegener's theory is ____ ; however, he died an outsider treated with ____ by the scientific establishment.**

    A. unsupported - approval

    B. dismissed - contempt

    C. accepted - approbation

    D. unchallenged - disdain

    E. unrivalled - reverence

22. **The revolution in art has not lost its steam; it ____ on as fiercely as ever.**

    A. trudges

    B. meanders

    C. edges

    D. ambles

    E. rages

23. Each occupation has its own ____ ; bankers, lawyers and computer professionals, for example, all use among themselves language which outsiders have difficulty following.

A. merits

B. disadvantages

C. rewards

D. jargon

E. problems

24. ____ by nature, Jones spoke very little even to his own family members.

A. garrulous

B. equivocal

C. taciturn

D. arrogant

E. gregarious

25. Biological clocks are of such ____ adaptive value to living organisms, that we would expect most organisms to ____ them.

A. clear - avoid

B. meager - evolve

C. significant - eschew

D. obvious - possess

E. ambivalent - develop

26. The peasants were the least ____ of all people, bound by tradition and ____ by superstitions.

A. free - fettered

B. enfranchised - rejected

C. enthralled - tied

D. pinioned - limited

E. conventional - encumbered

27. Many people at that time believed that spices help preserve food; however, Hall found that many marketed spices were ____ bacteria, moulds and yeasts.

A. devoid of

B. teeming with

C. improved by

D. destroyed by

E. active against

28. If there is nothing to absorb the energy of sound waves, they travel on ____ , but their intensity ____ as they travel further from their source.

A. erratically - mitigates

B. eternally - alleviates

C. forever - increases

D. steadily - stabilizes

E. indefinitely - diminishes

29. **The two artists differed markedly in their temperaments; Palmer was reserved and courteous, Frazer ____ and boastful.**

A. phlegmatic

B. choleric

C. constrained

D. tractable

E. stoic

30. **The intellectual flexibility inherent in a multicultural nation has been ____ in classrooms where emphasis on British-American literature has not reflected the cultural ____ of our country.**

A. eradicated - unanimity

B. encouraged - aspirations

C. stifled - diversity

D. thwarted - uniformity

E. inculcated - divide

IV. Paragraph Improvement (10 questions)

Instruction:

In this section, two draft passages are cited. For each passage, questions are set to test candidates' skills in improving the draft. The focus of the questions is on writing skills, not power of understanding.

Psaasge 1

(1)One of the big problems for geologists has always been determining the age of the fossils and other materials they find underground. (2)One method of wide usage is carbon-14 dating. (3)Carbon-14 dating is pretty straightforward. (4)Carbon-14 is a harmless radioactive material found in the Earth's atmosphere. (5) Animals and plants absorb carbon-14 into their tissues while they are alive. (6)After they die, however, they cease to absorb carbon-14, and the carbon-14 in their tissues begins to decay. (7)The process of decay that makes carbon-14 dating possible. (8)Scientists who study radioactive material like carbon-14 have developed the term "half-life" . (9)The "half-life" of a substance is the amount of time it takes for half of the atoms in a sample to decay. (10)Different radioactive materials have half-lives ranging from seconds to thousands of years. (11) Carbon-14 has a half-life of 5,730 years.

(12)A scientist named Willard Libby was the first person to use carbon-14 dating, in 1949. (13)Libby used the concept of "half-life" to determine that plant and animal remains containing half the amounts of carbon-14 expected in a living specimen were 5,730 years old. (14)Those remains containing less than half the expected amounts were older; the ones that contained more than half were younger.

Questions:

31. In context, which of the following is the best version of the underlined portion of sentence 2(reproduced below)?

One method of wide usage is carbon-14 dating.

A. (as it is now)

B. One method has been widely used, it

C. One widely used method

D. One method theuse of which has been wide

32. In context, which of the following is the best version of the underlined portion of sentences 3 and 4(reproduced below)?

Carbon-14 dating is pretty straightforward. Carbon-14 is a harmless radioactive material found in the Earth's atmosphere.

A. (as it is now)

B. is pretty straightforward. It

C. is pretty straightforward, because carbon-14

D. is pretty straightforward. However, carbon-14

33. In context, which of the following sentences would be best to insert between sentence 12 and sentence 13?

A. Libby was preoccupied with the problem of harmful radioactive waste.

B. He was studying a group of specimens that showed no signs of the presence of carbon-14.

C. His mentor, Dr. Frank McLean, had been the first geologist to employ the carbon-14 method.

D. Libby realized that the plant and animal remains he was studying contained traces of carbon-14.

34. The main purpose of paragraph 2 is to:

A. present a memorable example

B. summarize previous information

C. introduce an important concept

D. explain a puzzling contradiction

35. In context, what is the best way to deal with sentence 7 (reproduced below)?

The process of decay that makes carbon-14 dating possible.

A. Leave it as it is.

B. Delete "that"

C. Place a comma before "that"

D. Insert "is the thing" before "that"

Passage 2:

(1)Shoppers in the United States beware-there's a new way to buy groceries, and it's coming to a store near you. (2)As recently as 1999, only 6% of U.S. supermarkets had self-checkout lines. (3)By 2003 the number had risen to 38%. (4)And half of those supermarkets in the survey without self-checkout said that they planned to add the service. (5)Hardware giant Home Depot was among the first major retailers to use these machines. (6)I remember when I first took the self-checkout plunge. (7)It was a large chain supermarket near my office. (8)One day I noticed four large electronic units taking up prime space in a prominent corner of the store, beckoning shoppers with the promise of shorter lines. (9)I was attracted, not fearful. (10)It's so easy to let the cashier scan the groceries, check the coupons, take the money - was I up to the challenge of do-it-yourself. (11)I saw myself all elbows and thumbs, fumbling with my groceries and wallet as I battled the flashing computer screen. (12)Circumstances were driving me to face my fears. (13)When I at last approached the head of my line, the cashier announced it was time for her break.

Questions:

36. Which sentence should be deleted from the essay because it contains unrelated information?

A. Sentence 2

B. Sentence 5

C. Sentence 7

D. Sentence 8

37. All of the following strategies are used by the writer of the passage EXCEPT:

A. statistical evidence

B. imaginative description

C. personal narration

D. direct quotation

38. In context, which of the following is the best phrase to insert at the beginning of sentence 12?

A. But as it turned out,

B. In addition,

C. The result was that

D. In most cases,

39. In context, which of the following is the best version of the underlined portion of sentences 6 and 7(reproduced below)?

I remember when I first took the self-checkout plunge. It was a large chain supermarket near my office.

A. (as it is now)

B. plunge. It happened that there was

C. plunge. It happened at

D. plunge, which was

40. **The best way to describe the relationship of paragraph 2 to paragraph 1 is that paragraph 2:**

A. uses an anecdote to illustrate the facts presented in paragraph 1

B. provides facts to support a claim made in paragraph 1

C. tells a story that contradicts a theory offered in paragraph 1

D. uses statistics to back up a point made in paragraph 1

- END OF PAPER -

CC-CRE-UE

文化會社出版社 **CULTURE CROSS LIMITED**

答題紙 ANSWER SHEET

請在此貼上電腦條碼
Please stick the barcode label here

(1) 考生編號 Candidate No.

(2) 考生姓名 Name of Candidate

(3) 考生簽署 Signature of Candidate

宜用H.B.鉛筆作答
You are advised to use H.B. Pencils

考生須依照下圖所示填畫答案：

23 A B C D E

錯填答案可使用潔淨膠擦將筆痕徹底擦去。
切勿摺皺此答題紙

Mark your answer as follows:

23 A B C D E

Wrong marks should be completely erased with a clean rubber.

DO NOT FOLD THIS SHEET

1	A	B	C	D	E	21	A	B	C	D	E
2	A	B	C	D	E	22	A	B	C	D	E
3	A	B	C	D	E	23	A	B	C	D	E
4	A	B	C	D	E	24	A	B	C	D	E
5	A	B	C	D	E	25	A	B	C	D	E
6	A	B	C	D	E	26	A	B	C	D	E
7	A	B	C	D	E	27	A	B	C	D	E
8	A	B	C	D	E	28	A	B	C	D	E
9	A	B	C	D	E	29	A	B	C	D	E
10	A	B	C	D	E	30	A	B	C	D	E
11	A	B	C	D	E	31	A	B	C	D	E
12	A	B	C	D	E	32	A	B	C	D	E
13	A	B	C	D	E	33	A	B	C	D	E
14	A	B	C	D	E	34	A	B	C	D	E
15	A	B	C	D	E	35	A	B	C	D	E
16	A	B	C	D	E	36	A	B	C	D	E
17	A	B	C	D	E	37	A	B	C	D	E
18	A	B	C	D	E	38	A	B	C	D	E
19	A	B	C	D	E	39	A	B	C	D	E
20	A	B	C	D	E	40	A	B	C	D	E

文化會社出版社
投考公務員 模擬試題王

英文運用
模擬試卷（二）

時間：四十分鐘

考生須知：

（一）細讀答題紙上的指示。宣布開考後，考生須首先於適當位置貼上電腦條碼及填上各項所需資料。宣布停筆後，考生不會獲得額外時間貼上電腦條碼。

（二）試場主任宣布開卷後，考生請檢查試題冊及確定試題冊內的試題。最後會有「**全卷完**」的字眼。

（三）本試卷各題佔分相等。

（四）**本試卷全部試題均須回答**。為便於修正答案，考生宜用HB鉛筆把答案填畫在答題紙上。錯誤答案可用潔淨膠擦將筆痕徹底擦去。考生須清楚填畫答案，否則會因答案未能被辨認而失分。

（五）每題只可填畫**一個**答案。如填劃超過一個答案，該題將**不獲評分**。

（六）答案錯誤，不另扣分。

（七）未經許可，請勿打開試題冊。

CC-CRE-UE

I. Comprehension (10 questions)

Instruction:

This section aims to test candidates' ability to comprehend a written text. A prose passage of non-technical background is cited. Candidates are required to exercise skills in deciding on the gist, identifying main points, drawing inferences, distinguishing facts from opinion, interpreting figurative language, etc.

Passage:

Did you know that humans aren't the only species that use language? Bees communicate by dancing. Whales talk to each other by singing. And some apes talk to humans by using American Sign Language.

Meet Koko: a female gorilla born at the San Francisco Zoo on July 4th, 1971. Koko learned sign language from her trainer, Dr. Penny Patterson. Patterson began teaching sign language to Koko in 1972, when Koko was one year old. Koko must have been a good student, because two years later she moved onto the Stanford University campus with Dr. Patterson. Koko continued to learn on the campus until 1976. That's when she began living full-time with Patterson's group, the Gorilla Foundation. Patterson and Koko's relationship has blossomed ever since.

Dr. Patterson says that Koko has mastered sign language. She says that Koko knows over 1,000 words, and that Koko makes up new words. For example, Koko didn't know the sign for ring, so she signed the words "finger" and "bracelet" . Dr. Patterson thinks that this shows meaningful and constructive use of language.

Not everyone agrees with Dr. Patterson. Some argue that apes like Koko do not understand the meaning of what they are doing. Skeptics say that these apes are just performing complex tricks. For example, if Koko points to an apple and signs red or apple, Dr. Patterson will give her an apple. They argue that Koko does not really know what the sign apple means. She only knows that that if she makes the right motion, one which Dr. Patterson has shown her, then she gets an apple. The debate is unresolved, but one thing is for certain: Koko is an extraordinary ape.

Sign language isn't the only unusual thing about Koko. She's also been a pet-owner. In 1983, at the age of 12, researchers said that Koko asked for a cat for Christmas. They gave Koko a stuffed cat. Koko was not happy. She did not play with it, and she continued to sign sad. So for her birthday in 1984, they let her pick a cat out of an abandoned liter. Koko picked a gray cat and named him "All Ball" . Dr. Patterson said that Koko loved and nurtured All Ball as though he were a baby gorilla. Sadly, All Ball got out of Koko's cage and was hit by a car. Patterson reported that Koko signed "Bad, sad, bad" and "Frown, cry, frown, sad" when she broke the news to her.

It seems like Patterson and Koko have a good relationship, but not everyone agrees with it. Some critics believe that Patterson is humanizing the ape. They believe that apes should be left in the most natural state possible. Even Dr. Patterson struggles with these feelings. When asked if her findings could be duplicated by another group of scientists, she said, "We don't think that it would be ethical to do again." She went on to argue that animals should not be kept in such unnatural circumstances. Nonetheless, Koko lives in her foundation today.

As for the future, Dr. Patterson and the Gorilla Foundation would love to get Koko to an ape preserve in Maui, but they are having trouble securing the land. So unless you have a few million dollars to spare, Koko's going to be spending her time in Woodland, California with Dr. Patterson. Koko probably doesn't mind that. If she moved to Hawaii, she'd have to give up her Facebook page and Twitter feed, and she's got like 50 thousand "likes" . Some may deny that she knows sign language, but nobody says that she doesn't know social networking.

Questions:

1. **Which best expresses the main idea of this article?**
   A. Bees, whales, and apes like Koko all use language to communicate.
   B. Koko uses sign language but some think it's just a trick.
   C. It is natural for gorillas and house cats to live together.
   D. If you want a lot of "likes" on Facebook, get a talking gorilla.

2. **Which best describes how the second paragraph is organized?**

A. Chronological order

B. Cause and effect

C. Compare and contrast

D. Problem and solution

3. **Which best expresses the author's purpose in writing the second paragraph?**

A. The author is describing the environment in which Koko lives.

B. The author is informing readers how Dr. Patterson developed her skills.

C. The author is persuading readers that Koko should be freed.

D. The author is telling readers about Koko and Dr. Patterson's background.

4. **Which happened last?**

A. Koko got a stuffed cat for Christmas.

B. Koko lost All Ball.

C. Koko began living with the Gorilla Foundation.

D. Dr. Patterson began teaching Koko to sign.

5. **Which statement would the author most likely agree with?**

   A. Koko has mastered sign language without a doubt.

   B. Everybody likes how Dr. Patterson has raised Koko.

   C. Koko doesn't really know sign language.

   D. Some people are troubled by how Koko was raised.

6. **Which best defines the word duplicated as it is used in the sixth paragraph?**

   A. To dispute a fact or disagree with someone

   B. To lie to someone or to fool them

   C. To copy or recreate something

   D. To be disproven through debate

7. **Which event happened first?**

   A. Koko moved onto the Stanford University campus.

   B. Koko picked All Ball out for her birthday.

   C. Koko began living with the Gorilla Foundation.

   D. Koko got a stuffed cat for Christmas.

8. **Which best describes the main idea of the sixth paragraph?**
   A. Dr. Patterson has treated Koko very cruelly.
   B. Dr. Patterson and Koko have a beautiful, pure, and unconflicted relationship.
   C. Some people think that Koko should not have been treated like a human.
   D. Some people are working very hard to prove that Dr. Patterson is wrong.

9. **Which statement would the author most likely disagree with?**
   A. Dr. Patterson has worked hard to teach Koko sign language.
   B. Some people think that Koko only signs to get food.
   C. The Gorilla Foundation would like to move Koko to an ape preserve.
   D. Dr. Patterson has no regrets about working with Koko.

10. **If a book were being written about Koko and All Ball, which title would best summarize their story?**
    A. Long Wanted, Short Lived: A Tale of Strong Loves Lost
    B. Happy Ending: The Gorilla Who Got What She Wanted
    C. A Tale of Two Kitties: A Stuffed Cat Versus a Real One
    D. Plushy Love: How A Gorilla Fell in Love with a Stuffed Cat

II. Error Identification (10 questions)

Instruction:

Knowledge on use of the language is tested through identification of language errors which may be lexical, grammatical or stylistic.

Questions:

11. **The ROV asked a handicapped person in a wheelchair to hail taxis and, in less than five minutes, a taxi came and got out to help.**

    A. ROV

    B. in a wheelchair

    C. hail taxis

    D. got out to help

    E. No error

12. **A job-seeker was rejected but the Personnel Manager who rang up with the bad news promised to help him looked for a job elsewhere.**

    A. the Personnel Manager

    B. with the bad news

    C. looked for

    D. elsewhere

    E. No error

13. A rich man is looking into possible ventures in Philippines to develop a choice piece of real estate in the reclaimed area of Manila.

    A. ventures
    B. in Philippines
    C. a choice piece
    D. the reclaimed area
    E. No error

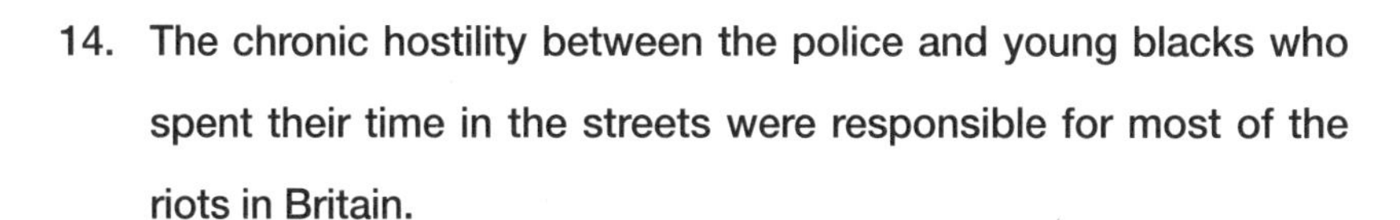

14. The chronic hostility between the police and young blacks who spent their time in the streets were responsible for most of the riots in Britain.

    A. blacks
    B. were
    C. most of
    D. in Britain
    E. No error

15. I am deeply disturbed by a common phenomena I observed in major cities, which, if unchecked, may flourish here.

    A. phenomena
    B. in major cities
    C. which
    D. if unchecked
    E. No error

16. If you believe you have what we are looking for and wish to embark on a satisfactory career in an organization that provides excellent growth and opportunities, submit your application immediately, give details of your personal particulars, qualifications and experience.

    A. what we are looking for

    B. embark on

    C. submit

    D. give

    E. No error

17. The Indonesian Foreign Minister said there should be conciliation, not confrontation, at the international conference on Cambodia with regard to iron out differences between the ASEAN and the Chinese views on how to resolve the conflict.

    A. not confrontation

    B. on Cambodia

    C. iron out

    D. how to resolve

    E. No error

18. The guided group discussion, as with other teaching methods, is designed to assist students in solving recognizing difficulties.

    A. guided
    B. as with
    C. is designed
    D. recognizing
    E. No error

19. In today's age of space exploration, traditional boundaries become less important, as man finds it necessary relying more and more on all the peoples of the world.

    A. today's
    B. man finds
    C. relying
    D. all the peoples
    E. No error

20. All too often, those students who seem highly motivated at the outset continue to exhibit such a behavior, while those who lack of this quality change their behavior very dramatically.

    A. All too often
    B. highly motivated
    C. such a behavior
    D. lack of
    E. No error

III. Sentence Completion (10 questions)

Instruction:

In this section, candidates are required to fill in the blanks with the best options given. The questions focus on grammatical use.

Questions:

21. **Unwilling to admit that they had been in error, the researchers tried to ____ their case with more data obtained from dubious sources.**

    A. ascertain

    B. buttress

    C. refute

    D. absolve

    E. dispute

22. **Archaeology is a poor profession; only ____ sums are available for excavating sites and even more ____ amounts for preserving the excavations.**

    A. paltry - meager

    B. miniscule - substantial

    C. average - augmented

    D. judicious - penurious

    E. modest - generous

23. The student was extremely foolhardy; he had the ____ to question the senior professor's judgment.

A. wisdom

B. temerity

C. interest

D. trepidation

E. condescension

24. The formerly ____ waters of the lake have been polluted so that the fish are no longer visible from the surface.

A. muddy

B. tranquil

C. stagnant

D. pellucid

E. rancid

25. After the accident, the nerves to her arm were damaged and so the muscles ____ through disuse.

A. atrophied

B. contracted

C. elongated

D. invigorated

E. dwindled

26. Some critics maintain that Tennyson's poetry is uneven, ranging from the ____ to the ____.

A. sublime - elevated

B. trite - inspired

C. vacuous - inane

D. succinct - laconic

E. sonorous - voluble

27. After grafting there is a ____ of lymphocytes in the lymph glands; the newly produced lymphocytes then move in to attack the foreign tissue.

A. diminution

B. proliferation

C. obliteration

D. paucity

E. attraction

28. One ____ the new scheme is that it might actually ____ just those applicants that it was intended to encourage.

A. highlight of - stimulate

B. feature of - attract

C. problem with - induce

D. attraction of - intimidate

E. drawback of - daunt

29. Corruption is ____ in our society; the integrity of even senior officials is ____ .

A. growing - unquestioned

B. endangered - disputed

C. pervasive - intact

D. rare - corrupted

E. rife - suspect

30. In their day to day decision making, many senior managers do not follow the rational model favored by orthodox management experts, but rather rely on intuitive processes that often appear ____ and ____.

A. cerebral - considered

B. heretical - judgmental

C. conscientious – logical

D. irrational - iconoclastic

E. capricious - deliberate

IV. Paragraph Improvement (10 questions)

Instruction:

In this section, two draft passages are cited. For each passage, questions are set to test candidates' skills in improving the draft. The focus of the questions is on writing skills, not power of understanding.

Passage 1

(1)Bizarre and mysterious events are things that happen all the time. (2)We have all had experiences that seemed supernatural, like unexpectedly meeting a long-lost friend when it was the day after dreaming about that friend. (3)Some people refer to these types of events as miracles. (4)Is there any scientific explanation for miracles?

(5)Professor Littlewood defined a miracle as an event with a probability of one in a million. (6)It is unlikely that you would ever see a miracle, right? (7)Well, Littlewood's Law of Miracles states that, for most people, miracles happen roughly once a month. (8)The proof of Littlewood's law is very simple. (9)He started by estimating that most people are active—at work or school, interacting with other people—at least 8 hours each day. (10)During this period, we see and hear things almost constantly; Littlewood estimated one new event per second. (11)These events add up fast because 1 per second for 8 hours equals 30,000 per day, or one million in one month.

(12)So if people are exposed to one million events per month, and a miracle is a one-in-a-million event, then we all experience a miracle once a month.

Questions:

31. What is the best way to deal with sentence 1(reproduced below)?

Bizarre and mysterious events are things that happen all the time.

A. Insert "to us" after "time"

B. Delete "and mysterious"

C. Change "are" to "could be"

D. Delete "are things that"

32. In context, what is the best version of the underlined portion of sentence 2(reproduced below)?

We have all had experiences that seemed supernatural, like unexpectedly meeting a long-lost friend when it was the day after dreaming about that friend.

A. (as it is now)

B. friend the day after

C. friend because it was the day after

D. friend who we saw the day after

33. In context, which of the following is the best phrase to insert at the beginning of sentence 6?

A. In any case,

B. Littlewood claimed that

C. At that rate,

D. Unless you are lucky,

34. Which of the following sentences would be most logical to insert before sentence 5?

A. A math professor named Littlewood has tried to explain miracles mathematically.

B. Some people believe that miracles cannot be explained by science.

C. Professor Littlewood taught mathematics at Cambridge University in England.

D. It would be helpful if we had a scientific law with which to explain miracles.

35. In context, which is the best version of the underlined portion of sentence 11 (reproduced below)?

These events add up fast because 1 per second for 8 hours equals 30,000 per day, or one million in one month.

A. (as it is now)

B. add up too fast for 1 per second for 8 hours to equal

C. add up fast: 1 per second for 8 hours equals

D. add up to 1 per second for 8 hours because it equals

Passage 2:

(1)Easter Island is a small island in the Pacific Ocean, over 2,000 miles off the coast. (2)Its distance from the coast of South America results in it being one of the most remote places in the world. (3)Easter Island is famous for its enormous stone statues of human heads and torsos. (4)Some of these statues are 70 feet tall and weigh over 200 tons.

(5)Scholars have puzzled over the origins of these statues, they are over 300 years old. (6)Archaeologists have located the stone quarry where the statues were built, but many of them are now located miles away. (7)How were such large objects moved? (8)Long thick ropes could have been made from tree bark and tied around the statues. (9) They could then have been dragged along the ground on sleds made of logs.

(10)This explanation seems to make sense. (11)European explorers arriving on Easter Island in the 1700s found no large trees or bushes anywhere on the island. (12)Ethnographer Kathleen Routledge visited Easter Island in 1914. (13)Where did the islanders get their ropes and sleds? (14)The answer lies in botany and its recent discoveries: Easter Island was once home to the world's largest type of palm tree.

Questions:

36. **In context, which is the best version of the underlined portion of sentences 1 and 2 (reproduced below)?**

    Easter Island is a small island in the Pacific Ocean, over 2,000 miles o the coast. Its distance from the coast of South America results in it being one of the most remote places in the world.

    A. (as it is now)

    B. off the coast of South America. It is considered

    C. off the coast. It is thus

    D. off the coast. As far as South America goes, it is

37. **Which sentence should be deleted from the essay because it contains unrelated information?**

    A. Sentence 4

    B. Sentence 6

    C. Sentence 9

    D. Sentence 12

38. **In context, which of the following revisions is necessary in sentence 5 (reproduced below)?**

    Scholars have puzzled over the origins of these statues, they are over 300 years old.

    A. Insert "that" before "they"

    B. Delete "they are"

    C. Change "they" to "which"

    D. Delete "statues"

39. Which of the following sentences would be most logical to insert before sentence 8?

A. The key seems to have been trees.

B. The islanders lacked sufficient organization.

C. Other groups utilized stone tools.

D. The only explanation is a supernatural one.

40. In context, which of the following is the best version of the underlined portion of sentences 10 and 11 (reproduced below)?

This explanation seems to make sense. European explorers arriving on Easter Island in the 1700s found no large trees or bushes anywhere on the island.

A. (as it is now)

B. seems to make sense, but European explorers

C. seems to make sense, and European explorers

D. seems to make sense, because European explorers

- END OF PAPER -

CC-CRE-UE

文化會社出版社 **CULTURE CROSS LIMITED**

答題紙 ANSWER SHEET

請在此貼上電腦條碼
Please stick the barcode label here

(1) 考生編號 Candidate No.

(2) 考生姓名 Name of Candidate

(3) 考生簽署 Signature of Candidate

宜用H.B.鉛筆作答
You are advised to use H.B. Pencils

考生須依照下圖所示填畫答案：

23 A B C D E

錯填答案可使用潔淨膠擦將筆痕徹底擦去。

切勿摺皺此答題紙

Mark your answer as follows:

23 A B C D E

Wrong marks should be completely erased with a clean rubber.

DO NOT FOLD THIS SHEET

1	A	B	C	D	E	21	A	B	C	D	E
2	A	B	C	D	E	22	A	B	C	D	E
3	A	B	C	D	E	23	A	B	C	D	E
4	A	B	C	D	E	24	A	B	C	D	E
5	A	B	C	D	E	25	A	B	C	D	E
6	A	B	C	D	E	26	A	B	C	D	E
7	A	B	C	D	E	27	A	B	C	D	E
8	A	B	C	D	E	28	A	B	C	D	E
9	A	B	C	D	E	29	A	B	C	D	E
10	A	B	C	D	E	30	A	B	C	D	E
11	A	B	C	D	E	31	A	B	C	D	E
12	A	B	C	D	E	32	A	B	C	D	E
13	A	B	C	D	E	33	A	B	C	D	E
14	A	B	C	D	E	34	A	B	C	D	E
15	A	B	C	D	E	35	A	B	C	D	E
16	A	B	C	D	E	36	A	B	C	D	E
17	A	B	C	D	E	37	A	B	C	D	E
18	A	B	C	D	E	38	A	B	C	D	E
19	A	B	C	D	E	39	A	B	C	D	E
20	A	B	C	D	E	40	A	B	C	D	E

文化會社出版社
投考公務員 模擬試題王

英文運用
模擬試卷（三）

時間：四十分鐘

考生須知：

（一）細讀答題紙上的指示。宣布開考後，考生須首先於適當位置貼上電腦條碼及填上各項所需資料。宣布停筆後，考生不會獲得額外時間貼上電腦條碼。

（二）試場主任宣布開卷後，考生請檢查試題冊及確定試題冊內的試題。最後會有「**全卷完**」的字眼。

（三）本試卷各題佔分相等。

（四）**本試卷全部試題均須回答**。為便於修正答案，考生宜用HB鉛筆把答案填畫在答題紙上。錯誤答案可用潔淨膠擦將筆痕徹底擦去。考生須清楚填畫答案，否則會因答案未能被辨認而失分。

（五）每題只可填畫**一個**答案。如填劃超過一個答案，該題將**不獲評分**。

（六）答案錯誤，不另扣分。

（七）未經許可，請勿打開試題冊。

CC-CRE-UE

I. Comprehension (10 questions)

Instruction:

This section aims to test candidates' ability to comprehend a written text. A prose passage of non-technical background is cited. Candidates are required to exercise skills in deciding on the gist, identifying main points, drawing inferences, distinguishing facts from opinion, interpreting figurative language, etc.

Passage:

Imagine that you're a fly. You're just zipping around the sky, looking for a place to rest, when you see nice pink leaf. That looks like a nice place to land. You think to yourself in your fly head. As you rest your feet on the leaf, you notice something strange. This leaf is hairy. You begin to make your move, but you trigger the plant's reflex. Snap! In one-tenth of a second, you are caught in the Venus flytrap. You will be digested in five to twelve days. Welcome to the world of carnivorous plants!

There are over a quarter of a millions plant species. Only 600 or so are carnivorous. We call them this because they attract, trap, and eat bugs. Like other plants, they get energy from the sun. But unlike other plants, they get their nutrients from their prey. Carnivorous plants live in bogs and places where the soil lacks nutrients. Most plants get nutrients from the soil. Carnivorous plants have turned to other sources.

The snap of the Venus flytrap is not the only way that plants eat bugs. Pitcher plants trick their prey into landing on them. They offer nectar bribes to the foolish insects that would take them. True to their name, pitcher plants have deep chambers. Their landing surface is slippery. They have inward pointing hairs, making it hard to escape. The fly lands on the pitcher plant to eat, but slips into a pit filled with digestive fluids and is eaten.

Then there're sundews. We call them sundews because they sparkle in the sun as if covered in morning dew. Of course, that sparkle is from something much more treacherous. It is a sweet goo called mucilage that bugs can't resist. Sundews create mucilage to attract bugs. As they fly in to eat, bugs become trapped in the very object of their desire. They soon exhaust themselves by trying to escape the mucilage. Or the sundew's tentacles, which respond to prey by curling around them, smother them. Bugs usually die in about 15 minutes. Then the plant dissolves its prey in enzymes and absorbs the nutrients.

Have you ever walked into trouble and found that you couldn't get out? So has every insect that has ever wandered into a corkscrew plant. Bugs love to investigate plants for nectar and food. Corkscrew plants have inviting stems. Curved hairs line the inside of these stems. These hairs allow insects to go up the stems, but not back. Going forward leads a chamber filled with digestive fluid, the plant's stomach. Bugs who wander into the corkscrew plant find that they are unable to escape. They must march to their own demise.

And then there are the bladderworts. They're about as nice as they sound. They live in water and float near the surface. Their traps are like small bladders hidden beneath the water. Only their flowers are visible from the surface. When bugs swim into the trigger hairs, the plant reacts. A trapdoor in the bladder opens up. The bladder sucks up the prey and the water surrounding it. A tenth of a second later, the bladder shuts again. The plant has trapped the prey. It releases digestive fluids. The prey will be digested within hours.

Carnivorous plants might sound tough, but they are difficult to keep at home. They are built to survive in places that other plants cannot. This specialization comes at a cost. They have a hard time adapting to other environments. Their strengths become weaknesses in rich soil. They depend on the harsh yet delicate environments in which they thrive. They are not so hardy after all. Still, there's something to be said about the power of life when one finds a plant that can survive in barren soil.

Questions:

1. **Which statement would the author most likely agree with?**

   A. There are too many species of carnivorous plants.

   B. There are too few plant species in the world.

   C. Only a small number of plants are carnivorous.

   D. A majority of plants are carnivorous.

2. **Which plant traps bugs in its stem and forces them to walk forward?**

   A. Corkscrew plants

   B. Sundews

   C. Bladderworts

   D. Pitcher plants

3. **Which of the following statements is false?**

   A. Carnivorous plants get their energy from eating bugs.

   B. Carnivorous plants do not get nutrients from the soil.

   C. Carnivorous plants get their energy from the sun.

   D. Carnivorous plants get their nutrients from eating bugs.

4. **Which event happens last when a sundew eats a meal?**

   A. The sundew creates mucilage.

   B. The sundew's tentacles curl in response to the prey.

   C. The bug is attracted to the mucilage.

   D. The sundew releases enzymes.

5. **Which best expresses the main idea of the third paragraph?**

   A. There are more types of carnivorous plants than the Venus fly trap.

   B. The pitcher plant tricks bugs into falling into its stomach.

   C. The Venus flytrap kills its prey in a various ways.

   D. Some plants attract bugs by offering them nectar.

6. **Which best defines the word treacherous as it is used in the fourth paragraph?**

   A. Something that provides nutrients.

   B. Something that is very bright.

   C. Something that tastes delicious.

   D. Something that has a hidden danger.

7. **Which best describes the over all text structure of the second paragraph?**

   A. Chronological order

   B. Compare and contrast

   C. Sequential order

   D. Spatial

8. **Which statement would the author most likely disagree with?**

   A. Carnivorous plants cannot thrive in rich soil.

   B. Bladderworts react quickly when their trigger hairs are bumped.

   C. Carnivorous plants are tough and can live in any environment.

   D. Bladderworts hide their traps just below the surface of the water.

9. Which best expresses the main idea of the last paragraph?

A. Carnivorous plants are not hard to take care of because they feed themselves.

B. Carnivorous plants are delicate because they need certain conditions to survive.

C. Carnivorous plants are difficult to keep at home, but you should keep trying.

D. Carnivorous plants are inspirational and they are interesting to watch and own.

10. Which title best expresses the author's main purpose in writing this text?

A. Watch Out! How To Avoid Being Eaten by Carnivorous Plants

B. At Risk: How You Can Help to Preserve Carnivorous Plants

C. Venus flytrap: Nature's Most Beautiful and Dangerous Plant

D. Fatal flowers: Plants That Kill Insects.

II. Error Identification (10 questions)

Instruction:

Knowledge on use of the language is tested through identification of language errors which may be lexical, grammatical or stylistic.

Questions:

11. **It seems that if students become aware of the fact that they are being evaluated, their discussion behavior tends to be affecting markedly.**

    A. become aware of

    B. being evaluated

    C. tends to be affecting

    D. markedly

    E. No error

12. **Some teachers have sought to involve students only superficially in the teacher-pupil planning, telling students that they play an active part in the process and then proceeding to oppose almost all suggestions that offered.**

    A. have sought to involve

    B. telling students

    C. in the process

    D. that offered

    E. No error

13. **The same individuals tend to make contributions, introduce them to the group in a given manner, and to voice either positive or negative reactions.**

    A. to make contributions

    B. introduce them

    C. in a given manner

    D. to voice

    E. No error

14. **The unusually dull student cannot meet the standards of the average student, and there is no sound reason that such standards should be imposed on him.**

    A. the average student

    B. sound

    C. that

    D. such standards

    E. No error

15. **Taxi drivers who overcharges or refuses to pick up passengers at the railway station, may have to contend with stiffer penalties, if caught.**

    A. who overcharges or refuses

    B. contend with

    C. stiffer penalties

    D. caught

    E. No error

16. **Fnes and suspension of the licenses of hardcore taxi drivers if they persist on being discourteous to the disabled were among several deterrent measures suggested.**

A. persist on

B. the disabled

C. were

D. suggested

E. No error

17. **Prince Sihanouk presented, in an open letter to the UN Conference participants, a three-points plan for his country. His plan included a dispatch to Cambodia of a UN-sponsored international peace-keeping force.**

A. the UN Conference participants

B. three-points

C. of

D. UN-sponsored

E. No error

18. The spirit of teamwork can be fostered in schools through working and playing in a group. Teachers, too, can contribute towards building up this spirit by being examples themselves of good teamworker.

    A. can be fostered
    B. in a group
    C. being examples
    D. teamworker
    E. No error

19. On his first day as prime minister, Mr. X wasted no time in calling up to his office 16 ministers, deputy ministers and parliamentary secretaries, to inform them their new portfolios.

    A. On his first day
    B. calling up to his office
    C. to inform them
    D. portfolios
    E. No error

20. Science is more than a collection of unrelated facts; to be meaningful and valuable, it must be arranged to show generalized facts.

    A. unrelated facts
    B. to be meaningful and valuable
    C. it
    D. generalized
    E. No error

III. Sentence Completion (10 questions)

Instruction:

In this section, candidates are required to fill in the blanks with the best options given. The questions focus on grammatical use.

Questions:

21. He was treated like a ____ and cast out from his community.

    A. ascetic

    B. prodigy

    C. prodigal

    D. pariah

    E. tyro

22. The teacher accused me of ____ because my essay was so similar to that of another student.

    A. procrastination

    B. plagiarism

    C. celerity

    D. confusion

    E. decorum

23. We live in a ____ age; everyone thinks that maximizing pleasure is the point of life.

A. ubiquitous

B. propitious

C. sporadic

D. corrupt

E. hedonistic

24. Thankfully the disease has gone into ____ ; it may not recur for many years.

A. treatment

B. sequestration

C. quarantine

D. remission

E. oblivion

25. People from all over the world are sent by their doctors to breathe the pure, ____ air in this mountain region.

A. invigorating

B. soporific

C. debilitating

D. insalubrious

E. aromatic

26. As were many colonial administrators, Gregory was ____ in his knowledge of the grammar of the local language, though his accent was ____ .

A. deficient - poor

B. competent - adequate

C. faultless - awful

D. well-versed - effective

E. erratic - eccentric

27. Though Adam Bede is presented to us by the author as ____ fiction, there are none of the life-like meanderings of the story of Amos Barton.

A. realistic

B. romantic

C. imaginative

D. educational

E. entertaining

28. There is a general ____ in the United States that our ethics are declining and that our moral standards are ____ .

A. feeling - normalizing

B. idea - futile

C. optimism - improving

D. complaint - deteriorating

E. outlook - escalating

29. Homo sapiens, the proud splitter of the atom, inventor of the electronic computer, ____ of the genetic code may be humbled by a lowly ____ of the sewers and soils - the microbe.

A. designer - inhabitant

B. discoverer - rodent

C. writer - organism

D. decipherer - denizen

E. author - purifier

30. After centuries of obscurity, this philosopher's thesis is enjoying a surprising ____ .

A. dismissal

B. remission

C. decimation

D. longevity

E. renaissance

IV. Paragraph Improvement (10 questions)

Instruction:

In this section, two draft passages are cited. For each passage, questions are set to test candidates' skills in improving the draft. The focus of the questions is on writing skills, not power of understanding.

Passage 1

(1)In the history of science, there are some discoveries that nobody can forget, one of them is when Copernicus proved that the Earth revolves around the Sun. (2)The scientists who made these groundbreaking discoveries must be among the smartest people of all time. (3)It's amazing that they could even come up with the ideas for their discoveries. (4)Much less go out and prove them. (5)But even the world's smartest people aren't perfect.

(6)Before Copernicus came along, people believed that the sun and the other planets circled around the Earth. (7)The models people used to show that the Earth was the center of the universe had stood the test of time: they were 1500 years old!. (8)It's hard to believe anyone could overturn 1500 years of scientific teaching. (9)I know I couldn't. (10)But astronomers today can see that Copernicus's actual calculations were wrong. (11)He was correct in general, but he had trouble with the details. (12)The same thing is true of John Dalton, who founded modern atomic theory. (13)The fact that these scientists are so celebrated, despite their flaws, just shows how groundbreaking their new ideas actually were. (14)Dalton's ideas were incredibly powerful and influential, but he used the wrong mathematical formulas.

Questions:

31. Which of the following is the best way to revise sentence 1(reproduced below)?

In the history of science, there are some discoveries that nobody can forget, one of them is when Copernicus proved that the Earth revolves around the Sun.

A. Change "are" to "were".

B. Insert "and" between "forget" and "one".

C. Change "one of them is" to "such as".

D. Insert "who" between "Copernicus" and "proved".

32. Of the following, which is the best way to revise and combine the underlined portions of sentences 3 and 4(reproduced below)?

It's amazing that they could even come up with the ideas for their discoveries. Much less go out and prove them.

A. (As it is now)

B. for their discoveries; much less

C. for their discoveries and much less

D. for their discoveries, much less

33. Which of the following sentences would be most logical to insert before sentence 6?

A. Take Copernicus, for example.

B. Copernicus was a Polish astronomer.

C. Many scientists led unhappy lives.

D. Famous scientists, politicians, and military leaders have all made mistakes.

34. In context, which is the best version of the underlined portion of sentence 7(reproduced below)?

The models people used to show that the Earth was the center of the universe had stood the test of time: they were 1500 years old!

A. (As it is now)

B. test of time, they were

C. test; their time was

D. test of time because they were

35. What should be done with sentence 9(reproduced below)?

I know I couldn't.

A. Leave it as it is.

B. Delete it.

C. Insert "Even if I were a brilliant astronomer," at the beginning.

D. Insert "that" between "know" and "I".

Passage 2

(1)Most people who visit museums, they are unaware that some of the paintings they are admiring are probably fakes. (2)It's hard to believe that forgers can trick the art experts. (3)But people have been doing it for years. (4)If a fake is considered to be good enough, the forger may be able to convince people that his work is actually a "lost masterpiece" of a famous artist!

(5)You would think that a trained art historian could take one look at a painting and tell who painted it and when. (6)In 1937 an art expert referred to a newly discovered painting they said was painted in the 17th century as "the masterpiece of Jan Vermeer" . (7)But in fact the painting was by a skilled Dutch forger named Hans van Meegeren. (8) Van Meegeren actually purchased a real 17th-century painting, and then he cleaned the paint off the canvas, and painted a new picture with a special paint he made himself. (9)The truth emerged when van Meegeren was arrested for trying to sell the painting to a foreign buyer. (10)There were many talented forgers in Germany. (11)Van Meegeren had to admit that the painting was fake to avoid going to jail for smuggling!

Questions:

36. **Which of the following is the best phrasing for the underlined portion of sentence 1 (reproduced below)?**

    Most people who visit museums, they are unaware that some of the paintings they are admiring are probably fakes.

    A. (As it is now)

    B. museums, unaware that

    C. museums are unaware that

    D. museums unaware, feel that

37. **Which sentence should be deleted from the passage because it is unnecessary?**

    A. Sentence 2

    B. Sentence 7

    C. Sentence 9

    D. Sentence 10

38. **In context, which of the following is the best way to phrase the underlined portion of sentence 4(reproduced below)?**

    If a fake is considered to be good enough, the forger may be able to convince people that his work is actually a "lost masterpiece" of a famous artist!

    A. (As it is now)

    B. If a fake is good enough

    C. If it is good enough

    D. If it is considered to be good enough

39. In the context of the passage, which of the following would fit most logically between sentences 5 and 6(reproduced below)?

You would think that a trained art historian could take one look at a painting and tell who painted it and when. In 1937 an art expert referred to a newly discovered painting they said was painted in the 17th century as "the masterpiece of Jan Vermeer."

A. Painting is relatively easy to learn, but extremely difficult to master.

B. Some art historians have spent years studying both the artistic and technical aspects of painting.

C. The passage of time takes quite a toll on a painting; the paint will crack and colors will fade.

D. Some of the so-called experts need to go back to school, however.

40. In context, which is the best version of the underlined portion of sentence 6(reproduced below)?

In 1937 an art expert referred to a newly discovered painting they said was painted in the 17th century as "the masterpiece of Jan Vermeer".

A. (As it is now)

B. people said to be

C. said to have been

D. said it was

- END OF PAPER -

CC-CRE-UE

文化會社出版社 **CULTURE CROSS LIMITED**

答題紙 ANSWER SHEET

請在此貼上電腦條碼
Please stick the barcode label here

(1) 考生編號 Candidate No.

(2) 考生姓名 Name of Candidate

(3) 考生簽署 Signature of Candidate

宜用H.B.鉛筆作答
You are advised to use H.B. Pencils

考生須依照下圖所示填畫答案：

23 A B C D E

錯填答案可使用潔淨膠擦將筆痕徹底擦去。
切勿摺皺此答題紙

Mark your answer as follows:

23 A B C D E

Wrong marks should be completely erased with a clean rubber.

DO NOT FOLD THIS SHEET

1	A	B	C	D	E	21	A	B	C	D	E
2	A	B	C	D	E	22	A	B	C	D	E
3	A	B	C	D	E	23	A	B	C	D	E
4	A	B	C	D	E	24	A	B	C	D	E
5	A	B	C	D	E	25	A	B	C	D	E
6	A	B	C	D	E	26	A	B	C	D	E
7	A	B	C	D	E	27	A	B	C	D	E
8	A	B	C	D	E	28	A	B	C	D	E
9	A	B	C	D	E	29	A	B	C	D	E
10	A	B	C	D	E	30	A	B	C	D	E
11	A	B	C	D	E	31	A	B	C	D	E
12	A	B	C	D	E	32	A	B	C	D	E
13	A	B	C	D	E	33	A	B	C	D	E
14	A	B	C	D	E	34	A	B	C	D	E
15	A	B	C	D	E	35	A	B	C	D	E
16	A	B	C	D	E	36	A	B	C	D	E
17	A	B	C	D	E	37	A	B	C	D	E
18	A	B	C	D	E	38	A	B	C	D	E
19	A	B	C	D	E	39	A	B	C	D	E
20	A	B	C	D	E	40	A	B	C	D	E

文化會社出版社
投考公務員 模擬試題王

英文運用
模擬試卷（四）

時間：四十分鐘

考生須知：

（一）細讀答題紙上的指示。宣布開考後，考生須首先於適當位置貼上電腦條碼及填上各項所需資料。宣布停筆後，考生不會獲得額外時間貼上電腦條碼。

（二）試場主任宣布開卷後，考生請檢查試題冊及確定試題冊內的試題。最後會有「**全卷完**」的字眼。

（三）本試卷各題佔分相等。

（四）**本試卷全部試題均須回答**。為便於修正答案，考生宜用HB鉛筆把答案填畫在答題紙上。錯誤答案可用潔淨膠擦將筆痕徹底擦去。考生須清楚填畫答案，否則會因答案未能被辨認而失分。

（五）每題只可填畫**一個**答案。如填劃超過一個答案，該題將**不獲評分**。

（六）答案錯誤，不另扣分。

（七）未經許可，請勿打開試題冊。

CC-CRE-UE

I. Comprehension (10 questions)

Instruction:

This section aims to test candidates' ability to comprehend a written text. A prose passage of non-technical background is cited. Candidates are required to exercise skills in deciding on the gist, identifying main points, drawing inferences, distinguishing facts from opinion, interpreting figurative language, etc.

Passage:

If you plan on going to Hawaii, don't bring any pets. Hawaiians are wary of letting in foreign animals. Your beloved Rex or fi-fi could spend up to 120 days in quarantine. They have strict rules for importing animals. They carefully screen all incoming pets. Who could blame them? They've had problems with new animals in the past.

The black rat was introduced to Hawaii in the 1780s. These ugly suckers originated in Asia, but they migrated to Europe in the 1st century. Since then they've snuck on European ships and voyaged the world with them. These rats carry many diseases including the plague. They are also good at surviving and tend to displace native species. That means that after they infest an area, there will be fewer birds and more black rats. Most people prefer living around birds.

Since their arrival in Hawaii, black rats have been pests. They've feasted on sea turtle eggs. They've eaten tree saplings, preventing trees from being reforested. And they've been a leading cause in the extinction of more than 70 species of Hawaiian birds. They love to climb trees to eat bird eggs. They also compete with forest birds for food, such as snails, insects, and seeds.

Perhaps more troubling, black rats threaten humans. They spread germs and incubate disease. They are a vector for more than 40 deadly illnesses. Some think that rat-borne diseases have killed more people than war in the last 1,000 years. Rats also eat our food. They eat more than 20% of the world's farmed food. And that's why the mongoose was brought to Hawaii.

During the mid 1800s, the Hawaiian sugar industry was thriving. Americans were just realizing that they loved sugar. Hawaii was pretty much the only place in America where one could grow sugarcane. But those filthy vermin were tearing up the fields. Black rats were destroying entire crops. What's a plantation owner to do? The answer is simple. Import an animal known to kill rats. What could go wrong with that? In 1883 plantation owners imported 72 mongooses and began breeding them.

People revere the mongoose in its homeland of India. They are often kept tame in Indian households. Mongooses feed on snakes, rats, and lizards, creatures that most people dislike. They are also cute and furry. And they kill deadly cobras. What's not to love? Sadly, India is a much different place than Hawaii.

When the mongooses got to Hawaii, they did not wipe out the rats as plantation owners hoped. Instead, they joined them in ravaging the birds, lizards, and small plants that were native to Hawaii. It's not that the mongooses became friends with the rats. They still ate a bunch of them. But mongooses are not too different from most other animals: they go for the easy meal. In Hawaii they had a choice. Pursue the elusive black rat or munch on turtle eggs while tanning on the beach. Most took the easy route.

Now Hawaii has two unwanted guests defacing the natural beauty. The Hawaiians have learned their lesson. Talks of bringing in mongoose-eating gorillas have been tabled. So don't get uptight when they don't welcome your cat Mittens with open arms. They're trying to maintain a delicate ecosystem here.

Questions:

1. **Based on the text, which best explains how black rats were introduced to Hawaii?**

   A. The native Hawaiians imported them to solve a problem with their crops.

   B. The Asians brought them to Hawaii when they first arrived.

   C. The Europeans brought them on their ships.

   D. The rats were able to swim to Hawaii from Asia.

2. **Which best defines the word originate as it was used in the second paragraph?**

   A. To come from a place

   B. To go to a place

   C. To become independent

   D. To wander the world

3. **Which event happened first?**

   A. The mongoose was introduced to Hawaii

   B. The black rat was introduced to Hawaii

   C. The black rat migrated to Europe

   D. Plantation owners bred mongooses

4. **Which statement would the author most likely disagree with?**

   A. Black rats threaten many creatures native to Hawaii.

   B. Mongooses threaten many creatures native to Hawaii.

   C. Mongooses were brought to Hawaii intentionally.

   D. The only reason people dislike rats is because they are ugly.

5. **Which best express the author's main purpose in writing this text?**

   A. To persuade readers to protect the endangered mongoose

   B. To describe the habits and hazards of the black rat

   C. To inform readers about species that have invaded Hawaii

   D. To entertain readers with tales of a mongoose's adventures

6. **Which best expresses the main idea of the sixth paragraph?**

   A. This paragraph is about Indian culture and wildlife.

   B. This paragraph is about the mongoose's role in Indian society.

   C. This paragraph is about the lifecycle of the mongoose.

   D. This paragraph is about how mongooses migrated to India.

7. **Which statement is false according to information in the text?**

   A. Rats eat lots of vegetation and crops.

   B. Mongooses eat sea turtle eggs.

   C. Rats climb trees and eat bird eggs.

   D. Mongooses have spread more than 40 diseases.

8. **Which best explains why plantation owners imported mongooses to Hawaii?**

   A. Mongooses eat rats.

   B. Mongooses are fuzzy and adorable.

   C. Mongooses make great household pets.

   D. Mongooses kill deadly cobras.

9. **Which best defines the word revere as it is used in the sixth paragraph?**

   A. To dislike someone or something

   B. To respect someone or something

   C. To hunt someone or something

   D. To get rid of someone or something

10. **Which title best expresses the main idea of this text?**

A. Travel Procedures: Getting in and out of Hawaii with Pets

B. Unwanted: The Journey of the Black Rat to Hawaii

C. Uncovered: What the Real Rikki-Tikki-Tavi is Like

D. Backfired: Solving Problems with Problems in Hawaii

II. Error Identification (10 questions)

Instruction:

Knowledge on use of the language is tested through identification of language errors which may be lexical, grammatical or stylistic.

Questions:

11. **X: One of us cleaned the room soon.**

Y: But it has already been cleaned, hasn't it?

A. cleaned

B. soon

C. has already been cleaned

D. hasn't it

E. No error

12. X: Are you waited for me, Cara?

Y: Yes, I am. Where were you a few minutes ago?

A. Are you waited for me

B. Cara

C. Yes, I am

D. Where were you

E. No error

13. X: Don't worry for such trifling things. They are not important.

Y: It is difficult not to worry. Trifling things may turn out to be important things some day.

A. Don't worry for

B. not important

C. not to worry

D. some day

E. No error

14. X: There may be something wrong with my watch. It stopped working without my knowledge.

Y: Why not adjust it? It is exactly 2:30pm on my watch.

A. something wrong with my watch

B. without my knowledge

C. Why not adjust it

D. on my watch

E. No error

15. X: Where is John? I have been looking for him.

Y: I haven't seen him either. Some say he has been on vacation from last week.

A. I have been looking for him

B. either

C. Some

D. from last week

E. No error

16. X: Which book do you prefer, Maria? This one or that one?

Y: I prefer this one than that one. How about you?

X: I prefer that one.

A. This one or that one

B. I prefer this one than that one

C. How about you

D. I prefer that one

E. No error

17. My friend and I decided to go to the concert last night. We got to there at 8:15pm. We were a few minutes late.

A. My friend and I

B. decided to go

C. We got to there at

D. a few minutes late

E. No error

18. **There were traffic jams on my way to the cinema. Besides, the bus came late. Fortunately, I arrived the cinema on time.**

A. on my way to

B. Besides,

C. I arrived

D. on time

E. No error

19. **As Sam has forgotten to bring his book with him, he has to share it with John, who is sitting besides him.**

A. has forgotten

B. with him

C. it

D. besides

E. No error

20. **Victor Klimov has shown that by shrinking the elements of a solar cell down to a few nanometres, or millions of a millimetre, each captured photon can be made to generate not one, but two or even more charge carriers.**

A. that by shrinking

B. down to

C. millions

D. can be made to

E. No error

III. Sentence Completion (10 questions)

Instruction:

In this section, candidates are required to fill in the blanks with the best options given. The questions focus on grammatical use.

Questions:

21. **Scrooge, in the famous novel by Dickens, was a ____ ; he hated the rest of mankind.**

    A. misanthrope

    B. hypochondriac

    C. philanthropist

    D. hedonist

    E. sybarite

22. **A businessman must widen his horizons; a ____ attitude will get you nowhere in this age of global communications.**

    A. moderate

    B. petrified

    C. parochial

    D. diversified

    E. comprehensive

23. Our bookshelves at home display a range of books on wide-ranging subjects and in many languages, reflecting the ____ tastes of our family members.

A. anomalous

B. limited

C. arcane

D. furtive

E. eclectic

24. Plastic bags are ____ symbols of consumer society; they are found wherever you travel.

A. rare

B. ephemeral

C. ubiquitous

D. fleeting

E. covert

25. Dr. Stuart needs to ____ his argument with more experimental data; as it stands his thesis is ____ .

A. support - profound

B. bolster - acceptable

C. refine - satisfactory

D. buttress - inadequate

E. define - succinct

26. After an initially warm reception by most reviewers and continued ____ by conservative thinkers, Bloom's work came under heavy fire.

A. criticism

B. endorsement

C. denigration

D. counterattack

E. refutation

27. Through the 19th Century, the classics of Western Civilization were considered to be the ____ of wisdom and culture, and an ____ person - by definition - knew them well.

A. foundation - average

B. epitome - uneducated

C. cornerstone - obtuse

D. font - ecclesiastical

E. repository - educated

28. In this biography we are given a glimpse of the young man ____ pursuing the path of the poet despite ____ and rejection slips.

A. doggedly - disappointment

B. tirelessly - encouragement

C. sporadically - awards

D. successfully - acclaim

E. unsuccessfully - failure

29. All European countries are seeking to diminish the check upon individual ____ which state examinations with their ____ growth have bought in their train.

A. rights - liberating

B. liberties - empowering

C. spontaneity - tyrannous

D. foibles - inevitable

E. creativity - soporific

30. In keeping with his own ____ in international diplomacy, Churchill proposed a personal meeting of heads of government, but the effort was doomed to failure, as the temper of the times was ____ .

A. ideas - pluralistic

B. predilections - inimical

C. aversions - hostile

D. impulses - amicable

E. maxims - salacious

IV. Paragraph Improvement (10 questions)

Instruction:

In this section, two draft passages are cited. For each passage, questions are set to test candidates' skills in improving the draft. The focus of the questions is on writing skills, not power of understanding.

Passage 1

(1)A significant problem all across our state is garbage. (2)Our landfills are full. (3)It seems that we must either find new sites for landfills or employ other methods of disposal, like incineration. (4)Unfortunately, there are drawbacks to every solution that they think of. (5)Polluted runoff water often results from landfills. (6)With incineration of trash, you get air pollution. (7)People are criticized for not wanting to live near a polluting waste disposal facility, but really, can you blame them?

(8)Recycling can be an effective solution, but owners of apartment complexes and other businesses complain that recycling adds to their expenses. (9)Local governments enjoy the benefits of taxes collected from business and industry. (10)They tend to shy away from pressuring such heavy contributors to recycle.

(11)Perhaps those of us being concerned should encourage debate about what other levels of government can do to solve the problems of waste disposal. (12)We should make a particular effort to cut down on the manufacture and use of things that will not decompose quickly. (13)Certainly we should press individuals, industries, and all levels of government to take responsible action while we can still see green grass and trees between the mountains of waste.

Questions:

31. Which of the following would fit most logically between sentences 1 and 2?

A. A sentence citing examples of states that have used up available landfills

B. A sentence citing examples of successful alternatives to landfills

C. A sentence citing the number of new landfills in the state

D. A sentence citing the average amount of trash disposed of annually by each person in the state

32. Which of the following is the best way to phrase the underlined portion of sentence 4(reproduced below)?

Unfortunately, there are drawbacks to every solution that they think of.

A. (as it is now)

B. that has been proposed

C. that they have previously come up with

D. to which there are proposals

33. Which of the following is the best way to revise and combine sentences 5 and 6(reproduced below)?

Polluted runo water often results from landfills. With incineration of trash, you get air pollution.

A. With landfills, polluted runoff water will result, and whereas with incineration of trash, you get air pollution.

B. While on the one hand are landfills and polluted runoff water, on the other hand you have air pollution in the case of incineration of trash.

C. Landfills often produce polluted runoff water, and trash incineration creates air pollution.

D. Landfills and incineration that produce water and air pollution.

34. If sentence 8 were rewritten to begin with the clause "Although recycling can be an effective solution," the next words would most logically be:

A. and owners of apartment complexes and other businesses complain

B. yet owners of apartment complexes and other businesses complain

C. owners of apartment complexes and other businesses complain

D. mostly owners of apartment complexes and other businesses are complaining

35. In context, which of the following is the best way to combine sentences 9 and 10?

A. Local governments enjoy the benefits of taxes collected from business and industry, as they tend to shy away from pressuring such heavy contributors to recycle.

B. Because local governments enjoy the benefits of taxes collected from business and industry, they tend to shy away from pressuring such heavy contributors to recycle.

C. However, local governments enjoy the benefits of taxes collected from business and industry, they tend to shy away from pressuring such heavy contributors to recycle.

D. In addition to enjoying the benefits of taxes collected from business and industry, local governments tend to shy away from pressuring business and industry into recycling.

Passage 2

(1)Some of the world's greatest scientists have been women, and most people still tend to think of science as a "man's game" . (2) There are probably many reasons that more men than women had fame as scientists. (3)Unequal access to educational opportunities is certainly one. (4)But sometimes the reason is plain old-fashioned dishonesty. (5)James Watson, Francis Crick, and Maurice Wilkins were awarded the Nobel Prize for the discovery by them of the double helix structure of the DNA molecule. (6)The discovery is seen by most as one of the greatest contributions to the modern history of biology. (7)One of the most important pieces of evidence used by

Watson and Crick to figure out this structure was an x-ray diffraction photograph that had been taken by a woman, Rosalind Franklin. (8) Scientists often build on the work of other scientists, but they usually do so openly. (9)Franklin's photograph was secretly shown to Watson by her colleague Maurice Wilkins. (10)Who never told her what he had done. (11)And then Watson, Crick, and Wilkins gave Nobel Prize lectures that contained 98 references to the work of other scientists, not citing a single one of Franklin's papers. (12)Of them only Wilkins in his speech making even a casual reference to her when he said she made some "very valuable contributions to the x-ray analysis".

Questions:

36. In context, which of the following is the best change to make to sentence 1?

A. Insert "As one can see" at the beginning.

B. Insert "In the field of genetics" at the beginning.

C. Insert "of course" after "and".

D. Insert "yet" after "and".

37. What is the best way to deal with sentence 2?

A. Omit it.

B. Switch it with sentence 1.

C. Change "had" to "have achieved".

D. Change "many" to "lots of".

38. Which of the following sentences is best inserted after sentence 3?

A. They think of science as a field in which men have been traditionally encouraged to participate.

B. The failure of the educational system to nurture young girls' interest in science is certainly another.

C. Some of the best-known names in science are those of men such as Galileo and Einstein.

D. The girls in my school are not given the same opportunities to study scientific subjects as the boys are.

39. In context, which of the following is the best way to express the underlined portion of sentence 5(reproduced below)?

James Watson, Francis Crick, and Maurice Wilkins were awarded the Nobel Prize for the discovery by them of the double helix structure of the DNA molecule.

A. (As it is now)

B. They were awarded the Nobel Prize for the discovery

C. Watson, Crick, and Wilkins were awarded the Nobel Prize for the discovery by them

D. James Watson, Francis Crick, and Maurice Wilkins were awarded the Nobel Prize for their discovery

40. In context, which of the following is the best version of sentences 9 and 10(reproduced below)?

Franklin's photograph was secretly shown to Watson by her colleague Maurice Wilkins. Who never told her what he had done.

A. (As it is now)

B. In this case, Franklin's colleague Maurice Wilkins secretly showed Watson her photograph without telling her.

C. Unfortunately, she did not know that her colleague Maurice Wilkins had secretly shown this photograph to Watson.

D. In fact, Franklin's colleague Maurice Wilkins never told her that he had shown Watson the photograph.

- END OF PAPER -

CC-CRE-UE

文化會社出版社 **CULTURE CROSS LIMITED**

答題紙 ANSWER SHEET

請在此貼上電腦條碼
Please stick the barcode label here

(1) 考生編號 Candidate No.

(2) 考生姓名 Name of Candidate

(3) 考生簽署 Signature of Candidate

宜用H.B.鉛筆作答
You are advised to use H.B. Pencils

考生須依照下圖所示填畫答案：

23 A B C D E

錯填答案可使用潔淨膠擦將筆痕徹底擦去。
切勿摺皺此答題紙

Mark your answer as follows:

23 A B C D E

Wrong marks should be completely erased with a clean rubber.

DO NOT FOLD THIS SHEET

No.						No.					
1	A	B	C	D	E	21	A	B	C	D	E
2	A	B	C	D	E	22	A	B	C	D	E
3	A	B	C	D	E	23	A	B	C	D	E
4	A	B	C	D	E	24	A	B	C	D	E
5	A	B	C	D	E	25	A	B	C	D	E
6	A	B	C	D	E	26	A	B	C	D	E
7	A	B	C	D	E	27	A	B	C	D	E
8	A	B	C	D	E	28	A	B	C	D	E
9	A	B	C	D	E	29	A	B	C	D	E
10	A	B	C	D	E	30	A	B	C	D	E
11	A	B	C	D	E	31	A	B	C	D	E
12	A	B	C	D	E	32	A	B	C	D	E
13	A	B	C	D	E	33	A	B	C	D	E
14	A	B	C	D	E	34	A	B	C	D	E
15	A	B	C	D	E	35	A	B	C	D	E
16	A	B	C	D	E	36	A	B	C	D	E
17	A	B	C	D	E	37	A	B	C	D	E
18	A	B	C	D	E	38	A	B	C	D	E
19	A	B	C	D	E	39	A	B	C	D	E
20	A	B	C	D	E	40	A	B	C	D	E

文化會社出版社
投考公務員 模擬試題王

英文運用
模擬試卷（五）

時間：四十分鐘

考生須知：

（一）細讀答題紙上的指示。宣布開考後，考生須首先於適當位置貼上電腦條碼及填上各項所需資料。宣布停筆後，考生不會獲得額外時間貼上電腦條碼。

（二）試場主任宣布開卷後，考生請檢查試題冊及確定試題冊內的試題。最後會有「**全卷完**」的字眼。

（三）本試卷各題佔分相等。

（四）**本試卷全部試題均須回答**。為便於修正答案，考生宜用HB鉛筆把答案填畫在答題紙上。錯誤答案可用潔淨膠擦將筆痕徹底擦去。考生須清楚填畫答案，否則會因答案未能被辨認而失分。

（五）每題只可填畫**一個**答案。如填劃超過一個答案，該題將**不獲評分**。

（六）答案錯誤，不另扣分。

（七）未經許可，請勿打開試題冊。

CC-CRE-UE

I. Comprehension (10 questions)

Instruction:

This section aims to test candidates' ability to comprehend a written text. A prose passage of non-technical background is cited. Candidates are required to exercise skills in deciding on the gist, identifying main points, drawing inferences, distinguishing facts from opinion, interpreting figurative language, etc.

Passage:

"Click!" That's the sound of safety. That's the sound of survival. That's the sound of a seat belt locking in place. Seat belts save lives and that's a fact. That's why I don't drive anywhere until mine is on tight. Choosing to wear your seat belt is a simple as choosing between life and death. Which one do you choose?

Think about it. When you're driving in a car, you may be going 60 MPH or faster. That car is zipping down the road. Then somebody ahead of you locks up his or her brakes. Your driver doesn't have time to stop. The car that you are in crashes. Your car was going 60 miles per hour. Now it has suddenly stopped. Your body, however, is still going 60 MPH. What's going to stop your body? Will it be the windshield or your seat belt? Every time that you get into a car you make that choice. I choose the seat belt.

Some people think that seat belts are uncool. They think that seat belts cramp their style, or that seat belts are uncomfortable. To them I say, what's more uncomfortable? Wearing a seat belt or flying through a car windshield? What's more uncool? Being safely anchored to a car, or skidding across the road in your jean shorts? Wearing a seat belt is both cooler and more comfortable than the alternatives.

Let's just take a closer look at your choices. If you are not wearing your seat belt, you can hop around the car and slide in and out of your seat easily. That sounds like a lot of fun. But, you are also more likely to die or suffer serious injuries. If you are wearing a seat belt, you have to stay in your seat. That's no fun. But, you are much more likely to walk away unharmed from a car accident. Hmmm... A small pleasure for a serious pain. That's a tough choice. I think that I'll avoid the serious pain.

How about giving money away? Do you like to give your money away? Probably not. And when you don't wear your seat belt, you are begging to give your money away. That's because kids are required to wear seat belts in every state in America. If you're riding in a car, and you don't have a seat belt on, the police can give you or your driver a ticket. Then you will have to give money to the city. I'd rather keep my money, but you can spend yours how you want.

Wearing a seat belt does not make you invincible. You can still get hurt or killed while wearing your seat belt. But wearing them has proven to be safer than driving without them. You are much less likely to be killed in a car wreck if you are wearing a seat belt. You are much less likely to get seriously injured if you are wearing one. So why not take the safer way? Why not go the way that has been proven to result in fewer deaths? You do want to live, don't you?

Questions:

1. **Which title best expresses the main idea of this text?**

   A. Car Accidents: Ways That We Can Prevent Them

   B. Slow Down: Save Lives By Driving Slower

   C. Seat Belts: Wear Them to Survive Any Wreck

   D. Why Not? Improve Your Odds with Seat Belts

2. **Which best expresses the author's main purpose in writing this text?**

   A. To inform readers about seat belt laws

   B. To persuade readers to wear seat belts

   C. To entertain readers with stories and jokes about seat belts

   D. To describe what car accidents are like without seat belts

3. **Which best describes the text structure in the fourth paragraph?**

   A. Compare and contrast

   B. Chronological order

   C. Sequential order

   D. Problem and solution

4. Which best defines the word alternatives as it is used in the third paragraph?

A. Being safe

B. Being unsafe

C. Other choices

D. Driving fast

5. Which best expresses the main idea of the fifth paragraph?

A. Seat belts are a waste of money.

B. People don't like to give money away.

C. Not wearing a seat belt may cost you.

D. Seat belt laws save lives.

6. Which best defines the word invincible as it is used in the last paragraph?

A. Uncool

B. Difficult or impossible to see

C. Glow-in-the-dark

D. Unable to be harmed

7. Which statement would the author most likely agree with?

A. Being safe is more important than being cool.

B. Moving freely around a car is worth the risks.

C. Seat belts will keep you safe in any car accident.

D. You should be most concerned with your comfort.

8. Which argument is not made by the author?

A. Not wearing a seat belt can be expensive.

B. Penalties for not wearing a seat belt should increase.

C. Seat belts keep you from flying through the windshield.

D. Wearing a seat belt is cooler than suffering an injury.

9. Which statement would the author most likely disagree with?

A. Seat belts save lives.

B. Every state in America has seat belt laws.

C. You shouldn't drive anywhere until you are wearing your seat belt.

D. Seat belts increase your chances of being injured in a car wreck.

10. Which best explains why the author starts his essay with the word click?

A. He is trying to scare readers.

B. He is trying to get the reader's attention.

C. He is trying to remind readers how seat belts sound when clasped.

D. He is trying to describe what it's like to ride in a car.

II. Error Identification (10 questions)

Instruction:

Knowledge on use of the language is tested through identification of language errors which may be lexical, grammatical or stylistic.

Questions:

11. **All science students must listen to instructions attentively. Any negligence or carelessness on their part will lead into great danger.**

    A. listen to instructions

    B. attentively

    C. on their part

    D. will lead into great danger

    E. No error

12. **In a small country like Singapore, all must work hard in order to serve for the people in future.**

    A. like Singapore

    B. all

    C. serve for

    D. in future

    E. No error

13. In my opinion most boys like to play outdoor games. Girls, on the other hand, prefer playing toys.

A. In my opinion

B. to play outdoor games

C. on the other hand

D. prefer playing toys

E. No error

14. Because of having robbed the shop, Alvin was arrested and putting into prison for a period of two months. He was set free earlier for good conduct.

A. having robbed the shop

B. putting into prison

C. for a period

D. for good conduct

E. No error

15. X: Helen, you look sad. What's the matter with you?

Y: Nothing. But my beloved hen died after it had lay its last egg.

A. you look sad

B. What's the matter

C. Nothing

D. it had lay its last egg

E. No error

16. **X: Can you tell me why he was absent yesterday?**

Y: I don't know why. But he told me that he is unwell. I hope he has recovered now.

A. why he was absent

B. I don't know why

C. he is unwell

D. he has recovered now

E. No error

17. **X: What did you see? Last night?**

Y: I saw something terrible. A thief killed a man, took the money, opening the door, and made off with the cash at high speed.

A. What did you see

B. something terrible

C. opening the door

D. made off with the cash at high speed

E. No error

18. **X: While Todd was reading the newspaper, his wife was sewing.**

Y: What do you mean? Didn't they did the same thing at that very moment?

A. was reading the newspaper

B. his wife was sewing

C. What do you mean

D. did the same thing

E. No error

19. **Some students at the back of the class were not attentive. They did something else instead of listening the lecture. They were in trouble later.**

A. at the back of the class

B. did something else

C. instead of listening the lecture

D. in trouble later

E. No error

20. **From a watch shop in North Bridge Road, a honest European bought a conventional watch at a very high price.**

A. in North Bridge Road

B. a honest European

C. a conventional watch

D. at a very high price

E. No error

III. Sentence Completion (10 questions)

Instruction:

In this section, candidates are required to fill in the blanks with the best options given. The questions focus on grammatical use.

Questions:

21. To reach Simonville, the traveller needs to drive with extreme caution along the ____ curves of the mountain road that climbs ____ to the summit.

    A. serpentine - steeply
    B. jagged - steadily
    C. gentle - precipitously
    D. shady - steadily
    E. hair-raising - languidly

22. The cricket match seemed ____ to our guests; they were used to watching sports in which the action is over in a couple of hours at the most.

    A. unintelligible
    B. inconsequential
    C. interminable
    D. implausible
    E. evanescent

23. **Our present accountant is most ____ ; unlike the previous _____ incumbent, he has never made a mistake in all the years that he has worked for the firm.**

A. unorthodox - heretical

B. dependable - assiduous

C. punctilious - painstaking

D. meticulous - unreliable

E. meticulous - reliable

24. **The refugee's poor grasp of English is hardly an _____ problem; she can attend classes and improve within a matter of months.**

A. implausible

B. insuperable

C. inconsequential

D. evocative

E. injudicious

25. **We appreciated his ____ summary of the situation; he wasted no words yet delineated his position most ____ .**

A. comprehensive : inadequately

B. succinct : direfully

C. cogent : persuasively

D. verbose : concisely

E. grandiloquent : eloquently

26. His musical tastes are certainly ____ ; he has recordings ranging from classical piano performances to rock concerts, jazz and even Chinese opera.

A. antediluvian

B. eclectic

C. harmonious

D. sonorous

E. dazzling

27. I cannot conclude this preface without ____ that an early and untimely death should have prevented Persius from giving a more finished appearance to his works.

A. rejoicing

B. lamenting

C. affirming

D. commenting

E. mentioning

28. Before his marriage the Duke had led an austere existence and now regarded the affectionate, somewhat ____ behavior of his young wife as simply ____ .

A. restrained - despicable

B. childish - elevating

C. playful - sublime

D. frivolous - puerile

E. unpleasant - delightful

29. Wilson ____ that human beings inherit a tendency to feel an affinity and awe for other living things, in the same way that we are ____ to be inquisitive or to protect our young at all costs.

A. argues - encouraged

B. maintains - trained

C. contends - predisposed

D. fears - taught

E. demurs - genetically programmed

30. The pond was a place of reek and corruption, of ____ smells and of oxygen-starved fish breathing through laboring gills.

A. fetid

B. evocative

C. dolorous

D. resonant

E. good

IV. Paragraph Improvement (10 questions)

Instruction:

In this section, two draft passages are cited. For each passage, questions are set to test candidates' skills in improving the draft. The focus of the questions is on writing skills, not power of understanding.

Passage 1

(1)People today have placed emphasis on the kinds of work that others do, it is wrong. (2)Suppose a woman says she is a doctor. (3)Immediately everyone assumes that she is a wonderful person, as if doctors were incapable of doing wrong. (4)However, if you say you're a carpenter or mechanic, some people think that you're not as smart as a doctor or a lawyer. (5)Can't someone just want to do this because he or she loves the work?

(6)Also, who decided that the person who does your taxes is more important than the person who makes sure that your house is warm or that your car runs ? (7)I know firsthand how frustrating it can be. (8)They think of you only in terms of your job. (9)I used to clean houses in the summer because the money was good; but yet all the people whose houses I cleaned seemed to assume that because I was vacuuming their carpets I did not deserve their respect. (10)One woman came into the bathroom while I was scrubbing the tub. (11) She kept asking me if I had any questions. (12)Did she want me to ask whether to scrub the tub counter-clockwise instead of clockwise? (13)Her attitude made me angry! (14)Once I read that the jobs people consider important have changed. (15)Carpenters used to be much more admired than doctors. (16)My point is, then, that who I want to be is much more important than what I want to be!

Questions:

31. Of the following, which is the best way to phrase sentence 1 (reproduced below)?

People today have placed emphasis on the kinds of work that others do, it is wrong.

A. (As it is now)

B. People today place too much emphasis on the kinds of work that others do.

C. What kinds of work others do is being placed too much emphasis on by people today.

D. The wrong kind of emphasis had been placed on the kinds of work others do today.

32. In context, which of the following is the best way to revise and combine the underlined portions of sentences 2 and 3 (reproduced below)?

Suppose a woman says she is a doctor. Immediately everyone assumes that she is a wonderful person, as if doctors were incapable of doing wrong.

A. Suppose a woman says she is a doctor, but immediately

B. If a woman says she is a doctor, for instance, immediately

C. When a woman says she is a doctor, however, immediately

D. Immediately, if they say, for example, she is a doctor,

33. In context, the phrase do this in sentence 5 would best be replaced by:

A. hold this particular opinion

B. resist temptation

C. ask someone for assistance

D. become a carpenter or a mechanic

34. Which of the following is the best way to revise and combine the underlined portions of sentences 7 and 8 (reproduced below)?

I know firsthand how frustrating it can be. They think of you only in terms of your job.

A. be; they - people, that is - think of you

B. be when they are thinking of one

C. be how people think of you

D. be when people think of you

35. **In context, the phrase but yet in sentence 9 would best be replaced by:**

A. incidentally,

B. however,

C. in fact,

D. in addition,

Passage 2

(1)Aristotle was a great philosopher and scientist. (2)Aristotle lived in Greece over 2300 years ago. (3)Aristotle was extraordinarily curious about the world around him. (4)He was also a master at figuring out how things worked. (5)Aristotle passed it on to his pupil Theophrastus.

(6)Theophrastus was famous among his contemporaries as the co-founder of the Lyceum, a school in Greece, he is best known today as "the father of botany" . (7)Botany is the branch of science dealing with plants.

(8)Two famous books he wrote were Natural History of Plants and Reasons for Vegetable Growth. (9)His books were translated from Greek into Latin in 1483-1800 years after he wrote them - they influenced thousands of readers.

(10)Theophrastus made accurate observations about all aspects of plant life, including plant structure, plant diseases, seed use, and medicinal properties. (11)He even described the complex process of plant reproduction correctly, hundreds of years before it was formally proven. (12)In 1694 Rudolph Jakob Camerarius used experiments to show how plants reproduced. (13)According to some accounts, Theophrastus did his research in a garden he maintained at his school which was called the Lyceum. (14)But Theophrastus also wrote about plants that grew only in other countries, which he heard about from returning soldiers. (15)By comparing these plants to plants he grew in his garden, Theophrastus established principles that are still true today.

Questions:

36. Which of the following is the best version of the underlined portion of sentence 1 and sentence 2 (reproduced below)?

Aristotle was a great philosopher and scientist. Aristotle lived in Greece over 2300 years ago.

A. philosopher and a scientist, living

B. philosopher and scientist who lived

C. philosopher, and, as a scientist, lived

D. philosopher and scientist; Aristotle lived

37. What would best replace "it" in sentence 5?

A. that

B. them

C. these traits

D. the world

38. What word should be inserted between "Greece," and "he" in sentence 6 (reproduced below)?

Theophrastus was famous among his contemporaries as the co-founder of the Lyceum, a school in Greece, he is best known today as "the father of botany."

A. and

B. but

C. for

D. thus

39. Which sentence should be inserted between sentence 8 and sentence 9?

A. Theophrastus's ideas had a lasting impact.

B. Theophrastus's books were instantly successful.

C. The first book is still studied today in botany classes.

D. They challenged the conclusions of Aristotle.

40. **Which revision appropriately shortens sentence 13(reproduced below)?**

According to some accounts, Theophrastus did his research in a garden he maintained at his school which was called the Lyceum.

A. Delete "his school which was called"

B. Delete "According to some accounts,"

C. Delete "in a garden he maintained"

D. Replace "According to some accounts" with "Therefore"

- END OF PAPER -

CC-CRE-UE

文化會社出版社 **CULTURE CROSS LIMITED**

答題紙 ANSWER SHEET

請在此貼上電腦條碼 Please stick the barcode label here

(1) 考生編號 Candidate No.

(2) 考生姓名 Name of Candidate

(3) 考生簽署 Signature of Candidate

宜用H.B.鉛筆作答
You are advised to use H.B. Pencils

考生須依照下圖所示填畫答案：

23 A B C D E

錯填答案可使用潔淨膠擦將筆痕徹底擦去。

切勿摺皺此答題紙

Mark your answer as follows:

23 A B C D E

Wrong marks should be completely erased with a clean rubber.

DO NOT FOLD THIS SHEET

1	A	B	C	D	E	21	A	B	C	D	E
2	A	B	C	D	E	22	A	B	C	D	E
3	A	B	C	D	E	23	A	B	C	D	E
4	A	B	C	D	E	24	A	B	C	D	E
5	A	B	C	D	E	25	A	B	C	D	E
6	A	B	C	D	E	26	A	B	C	D	E
7	A	B	C	D	E	27	A	B	C	D	E
8	A	B	C	D	E	28	A	B	C	D	E
9	A	B	C	D	E	29	A	B	C	D	E
10	A	B	C	D	E	30	A	B	C	D	E
11	A	B	C	D	E	31	A	B	C	D	E
12	A	B	C	D	E	32	A	B	C	D	E
13	A	B	C	D	E	33	A	B	C	D	E
14	A	B	C	D	E	34	A	B	C	D	E
15	A	B	C	D	E	35	A	B	C	D	E
16	A	B	C	D	E	36	A	B	C	D	E
17	A	B	C	D	E	37	A	B	C	D	E
18	A	B	C	D	E	38	A	B	C	D	E
19	A	B	C	D	E	39	A	B	C	D	E
20	A	B	C	D	E	40	A	B	C	D	E

文化會社出版社
投考公務員 模擬試題王

英文運用
模擬試卷（六）

時間：四十分鐘

考生須知：

（一）細讀答題紙上的指示。宣布開考後，考生須首先於適當位置貼上電腦條碼及填上各項所需資料。宣布停筆後，考生不會獲得額外時間貼上電腦條碼。

（二）試場主任宣布開卷後，考生請檢查試題冊及確定試題冊內的試題。最後會有「**全卷完**」的字眼。

（三）本試卷各題佔分相等。

（四）**本試卷全部試題均須回答**。為便於修正答案，考生宜用HB鉛筆把答案填畫在答題紙上。錯誤答案可用潔淨膠擦將筆痕徹底擦去。考生須清楚填畫答案，否則會因答案未能被辨認而失分。

（五）每題只可填畫**一個**答案。如填劃超過一個答案，該題將**不獲評分**。

（六）答案錯誤，不另扣分。

（七）未經許可，請勿打開試題冊。

CC-CRE-UE

I. Comprehension (10 questions)

Instruction:

This section aims to test candidates' ability to comprehend a written text. A prose passage of non-technical background is cited. Candidates are required to exercise skills in deciding on the gist, identifying main points, drawing inferences, distinguishing facts from opinion, interpreting figurative language, etc.

Passage:

Did you know that some people don't do their reading assignments? It's shocking, but it's true. Some students don't even read short texts that they are assigned in class. There are many reasons for this. They may be distracted or bored. They may be unwilling to focus. They may be unconfident readers. Whatever the reason, it has to stop today. Here's why.

Reading stimulates your mind. It is like a workout for your brain. When people get old, their muscles begin to deteriorate. They get weaker and their strength leaves them. Exercise can prevent this loss. The same thing happens to people's brains when they get older. Brain power and speed decline with age. Reading strengthens your brain and prevents these declines.

You can benefit from reading in the near-term too. Reading provides knowledge. Knowledge is power. Therefore, reading can make you a more powerful person. You can learn to do new things by reading. Do you want to make video games? Do you want to design clothing? Reading can teach you all this and more. But you have to get good at reading, and the only way to get good at something is to practice.

Read everything that you can at school, regardless of whether you find it interesting. Reading expands your vocabulary. Even a "boring" text can teach you new words. Having a larger vocabulary will help you better express yourself. You will be able to speak, write, and think more intelligently. What's boring about that?

Do not just discount a text because it is unfamiliar to you. Each time you read, you are exposed to new ideas and perspectives. Reading can change the way that you understand the world. It can give you a broader perspective on things. It can make you worldlier. You can learn how people live in far away places. You can learn about cultures different from your own.

Reading is good for your state of mind. It has a calming effect. It can lower your stress levels and help you relax. You can escape from your troubles for a moment when you read, and it's a positive escape. The benefits of reading far outweigh those of acting like a doofus. So do yourself a favor: the next time you get a reading assignment, take as much as you can from it. Squeeze it for every drop of knowledge that it contains. Then move on to the next one.

Questions:

1. **Which best expresses the main idea of the second paragraph?**

   A. Reading is exciting.

   B. Reading strengthens your mind.

   C. Age affects the body in many ways.

   D. Working out keeps your body in shape.

2. **Why does the author think that you should read books that are boring?**

   A. You will eventually grow to love them if you read them enough.

   B. You will get better grades in reading class.

   C. You will make your teacher very happy.

   D. You will learn new words.

3. **Which best expresses the main idea of the third paragraph?**

   A. Reading can benefit you.

   B. You can learn to program video games or design clothing by reading.

   C. You can learn amazing things and become a better person by reading.

   D. Knowledge is power.

4. **Which is not a reason given by the author why students fail to complete reading assignments?**

   A. Students may be bored.

   B. Students may be distracted.

   C. Students may be unwilling to focus.

   D. Students may be tired.

5. **Which best expresses the author's main purpose in writing this text?**

   A. He is trying to persuade students to do their reading work.

   B. He is teaching people how to become better readers.

   C. He is explaining why people don't do their reading work.

   D. He is entertaining readers with facts about the mind and body.

6. **Which best describes the author's tone in the first three sentences?**

   A. Surprised

   B. Sarcastic

   C. Informative

   D. Irate

7. Which of the following is not one of the author's main points?

A. Reading broadens your perspective and makes you a better person.

B. Reading is a relaxing activity with positive mental side effects.

C. Reading helps you perform on tests and get into selective schools.

D. Reading keeps your mind in shape and prevents losses due to age.

8. Which is not one of the author's arguments in the fifth paragraph?

A. Reading gives you a broader perspective on the world.

B. Reading changes the way that you understand the world.

C. Reading helps prepare you for your job in the real world.

D. Reading teaches you about distant lands and cultures.

9. Why does the author believe that reading is good for your mind state?

A. It has a calming effect.

B. It can lower your stress levels.

C. It can help you relax.

D. All of these

10. **Which title best expresses the main idea of this text?**

    A. Reading: Good for the Mind in Many Ways

    B. Reading: The Key to a Successful Academic Future

    C. Reading: Improve Your Vocabulary While Being Entertained

    D. Reading: The Best Way to Improve Your Writing Skills

II. Error Identification (10 questions)

Instruction:

Knowledge on use of the language is tested through identification of language errors which may be lexical, grammatical or stylistic.

Questions:

11. **Please sitting down. The manager is on the phone. He says he will probably see you in a few minutes' time.**

    A. Please sitting down

    B. on the phone

    C. he will probably

    D. in a few minutes' time

    E. No error

12. **Some committee members disagreed with the chairman. They attacked on his illogical thinking. But the majority of them were not against him. Instead, they backed him up.**

    A. disagreed with the chairman

    B. They attacked on

    C. were not against him

    D. Instead, they backed him up

    E. No error

13. **We paid a visit to Australia last month. We spend not only some time in the urban area but also went to the countryside from time to time.**

    A. paid a visit to

    B. We spend not only some time

    C. in the urban area

    D. from time to time

    E. No error

14. **Please send me an application form together with the latest prospectus. If possible, also give me other relevant informations.**

    A. send me an application form

    B. the latest prospectus

    C. If possible

    D. other relevant informations

    E. No error

15. **A person should always keep one's word if he hopes to be trusted. On the other hand, nobody will have any respect for him if he breaks his promises too often.**

A. keep one's word

B. to be trusted

C. On the other hand

D. if he breaks his promises too often

E. No error

16. **X: When did the workers start working?**

Y: In the morning at about 8:30. And they went home after they had been finishing their work.

A. the workers start working

B. at about 8:30

C. And they went home

D. they had been finishing their work

E. No error

17. **X: Has your husband be to Hong Kong before?**

Y: I think he hasn't. He simply has no time.

A. Has your husband be

B. before

C. I think he hasn't

D. He simply has no time

E. No error

18. **X: Charlie and his family are living here for many years. Do you think so?**

Y: Yes, I certainly think so, but I'm not sure how many years.

A. are living here

B. so

C. I certainly think so

D. how many years

E. No error

19. **I'm enclosing a 500-dollars check in payment of the books ordered. Please acknowledge receipt at your earliest convenience.**

A. a 500-dollars check

B. ordered

C. acknowledge receipt

D. at your earliest convenience

E. No error

20. **There was a flood last month. Some people climbed up the house's roof. They were not in danger. But, as bad luck would have it, a few old men were drowned.**

A. the house's roof

B. in danger

C. as bad luck would have it

D. were drowned

E. No error

III. Sentence Completion (10 questions)

Instruction:

In this section, candidates are required to fill in the blanks with the best options given. The questions focus on grammatical use.

Question:

21. The crew of the air balloon ____ the sand bags to help the balloon rise over the hill.

    A. capsized

    B. jettisoned

    C. salvaged

    D. augmented

    E. enumerated

22. We were not fooled by his ____ arguments; his plan was obviously ____ .

    A. cogent - brilliant

    B. hackneyed - banal

    C. convoluted - labyrinthine

    D. specious - untenable

    E. lucid - intelligible

23. Hawkins is ____ in his field; no other contemporary scientist commands the same respect.

A. disparaged

B. ignominious

C. obsolete

D. anachronistic

E. preeminent

24. The model paraded in front of the celebrities with ____ ; it was impossible to tell that this was her first assignment.

A. panache

B. opprobrium

C. shame

D. trepidation

E. terror

25. The term lead pencil is a ____ ; pencils are filled with graphite not lead.

A. misnomer

B. misdemeanor

C. peccadillo

D. euphemism

E. metaphor

26. The ____ weather forced us to stay indoors.

A. enticing

B. glorious

C. restorative

D. inclement

E. congenial

27. It will be hard to ____ Leonid now that you have so ____ him.

A. pacify - soothed

B. mollify - incensed

C. antagonize - irritated

D. anger - ruffled

E. subdue - subjugated

28. The lectures on quantum physics were invariably ____ ; the lecturer ____ his ill-prepared material in a manner guaranteed to send even the most ardent student to sleep.

A. stimulating - delivered

B. pedestrian - enthused about

C. soporific - droned

D. scintillating - intoned

E. arcane - marshaled

29. **Edward was understandably upset that he had lost the position, but he was ____ by the conviction that he had done nothing to ____ the dismissal.**

   A. consoled - merit

   B. warmed - avoid

   C. comforted - mar

   D. miffed - delay

   E. saddened - earn

30. **Elinor ____ to counteract her negative feelings, but only succeeded in ____ them.**

   A. tried - allaying

   B. hoped - mitigating

   C. desired - ameliorating

   D. hesitated - deprecating

   E. endeavoured - intensifying

IV. Paragraph Improvement (10 questions)

Instruction:

In this section, two draft passages are cited. For each passage, questions are set to test candidates' skills in improving the draft. The focus of the questions is on writing skills, not power of understanding.

Passage 1:

(1)The early history of astronomy was full of misunderstandings. (2) Some of them were funny, it's like the controversy of the "canali" on Mars. (3)In the late 1800's an Italian astronomer named Giovanni Schiaparell studied Mars. (4)He had a high-powered telescope that hused to look at Mars. (5)Schiaparelli thought he saw channels criss-crossing the planet's surface. (6)He was intrigued: perhaps these channels were evidence that Marhad great flowing rivers like the Earth. (7)Schiaparelli made charts of the surface of Mars and labeled it with the Italian word "canali" .

(8)Unfortunately, "canali" can be translated into English as either "channels" or "canals" . (9)Channels and canals are two different things because channels are form naturally by water, while canals are constructed by people. (10)Some people translated "canali" as "canals" , word began to spread that the lines Schiaparelli saw through his telescope were actually canals that had been built by intelligent beings. (11)One of them was an amateur astronomer named Percival Lowell. (12)He wrote a serie of best-selling books. (13) In these books Lowell publicized the notion that these "canals" were built by Martian farmers who understood irrigation.

(14)In 1965 a U.S. spacecraft flying close to the surface of Mars sent back conclusive pictures. (15)There are no prominent channels anywhere on the planet. (16)Lowell and Schiaparelli saw what they wanted to see. (17)Lowell was wrong, of course, but so was Schiaparelli.

Questions:

31. **Which is the best version of the underlined portion of sentence 2(reproduced below)?**

    Some of them were funny, it's like the controversy of the "canali" on Mars.

    A. (as it is now)

    B. funny; it's like

    C. funny, like

    D. funny, even

32. **Which is the best way to combine sentences 3 and 4(reproduced below)?**

    In the late 1800's an Italian astronomer named Giovanni Schiaparelli studied Mars. He had a high-powered telescope that he used to look at Mars.

    A. In the late 1800's an Italian astronomer named Giovanni Schiaparelli studied Mars by a high-powered telescope.

    B. In the late 1800's an Italian astronomer named Giovanni Schiaparelli studied Mars with a high-powered telescope that he used to look at Mars.

    C. In the late 1800's an Italian astronomer named Giovanni Schiaparelli studied Mars, he had a high-powered telescope that he used.

    D. In the late 1800's an Italian astronomer named Giovanni Schiaparelli used a high-powered telescope to study Mars.

33. **Which word would be best to insert at the beginning of sentence 10(reproduced below)?**

Some people translated "canali" as "canals" , word began to spread that the lines Schiaparelli saw through his telescope were actually canals that had been built by intelligent beings.

A. Because

B. However

C. If

D. Although

34. **What is the best version of the underlined portion of sentence 11(reproduced below)?**

One of them was an amateur astronomer named Percival Lowell.

A. (As it is now)

B. One of the most intelligent was

C. This idea was popularized by

D. It was spread even more by someone else,

35. **What is the best way to combine sentences 12 and 13(reproduced below)?**

He wrote a series of bestselling books. In these books Lowell publicized the notion that these "canals" were built by Martian farmers who understood irrigation.

A. In a series of bestselling books, Lowell publicized the notion that these "canals" were built by Martian farmers who understood irrigation.

B. He wrote a series of books that was a bestseller and publicized the notion that these "canals" were built by Martian farmers who understood irrigation.

C. His books that were bestsellers publicized the notion that these "canals" were built by Martian farmers who understood irrigation.

D. He wrote a series of bestselling books; Lowell publicized the notion that these "canals" were built by Martian farmers who understood irrigation.

Passage 2:

(1)Not many children leave elementary school and they have not heard of Pocahontas' heroic rescue of John Smith from her own people, the Powhatans. (2)Generations of Americans have learned the story of a courageous Indian princess who threw herself between the Virginia colonist and the clubs raised to end his life. (3)The captive himself reported the incident. (4)According to that report, Pocahontas held his head in her arms and laid her own upon his to save him from death.

(5)But can Smith's account be trusted? (6)Probably it cannot, say several historians interested in dispelling myths about Pocahontas. (7) According to these experts, in his eagerness to find patrons for future expeditions, Smith changed the facts in order to enhance his image. (8) Portraying himself as the object of a royal princess' devotion may have merely been a good public relations ploy. (9)Research into Powhatan culture suggests that what Smith described as an execution might have been merely a ritual display of strength. (10)Smith may have been a character in a drama in which even Pocahontas was playing a role.

(11)As ambassador from the Powhatans to the Jamestown settlers, Pocahontas headed off confrontations between mutually suspicious parties. (12)Later, after her marriage to colonist John Rolfe, Pocahontas traveled to England, where her diplomacy played a large part in gaining support for the Virginia Company.

Questions:

36. What is the best way to deal with sentence 1(reproduced below)?

Not many children leave elementary school and they have not heard of Pocahontas' heroic rescue of John Smith from her own people, the Powhatans.

A. Leave it as it is.

B. Switch its position with that of sentence 2.

C. Change “leave” to “have left”.

D. Change “and they have not heard” to “without having heard”.

37. In context, which of the following is the best way to revise the underlined wording in order to combine sentences 3 and 4?

The captive himself reported the incident. According to that report, Pocahontas held his head in her arms and laid her own upon his to save him from death.

A. The captive himself reported the incident, according to which

B. According to the captive's own report of the incident,

C. Consequently, the captive himself reports that

D. It seems that in the captive's report of the incident he says that

38. Which of the following phrases is the best to insert at the beginning of sentence 10 to link it to sentence 9?

A. Far from being in mortal danger,

B. If what he says is credible,

C. What grade school history never told you is this:

D. They were just performing a ritual, and

39. Which of the following best describes the relationship between sentences 9 and 10?

A. Sentence 10 concludes that the theory mentioned in sentence 9 is wrong.

B. Sentence 10 adds to information reported in sentence 9.

C. Sentence 10 provides an example to illustrate an idea presented in sentence 9.

D. Sentence 10 poses an argument that contradicts the point made in sentence 9.

40. Which of the following would be the best sentence to insert before sentence 11 to introduce the third paragraph?

A. It is crucial to consider the political successes as well as the shortcomings of Pocahontas.

B. The Pocahontas of legend is the most interesting, but the historical Pocahontas is more believable.

C. If legend has overemphasized the bravery of Pocahontas, it has underplayed her political talents.

D. To really know Pocahontas, we must get beyond myth and legend to the real facts about her private life.

- END OF PAPER -

CC-CRE-UE

文化會社出版社 **CULTURE CROSS LIMITED**

答題紙 ANSWER SHEET

請在此貼上電腦條碼
Please stick the barcode label here

(1) 考生編號 Candidate No.

(2) 考生姓名 Name of Candidate

(3) 考生簽署 Signature of Candidate

宜用H.B.鉛筆作答
You are advised to use H.B. Pencils

考生須依照下圖所示填畫答案：

23 A B C D E

錯填答案可使用潔淨膠擦將筆痕徹底擦去。
切勿摺皺此答題紙

Mark your answer as follows:

23 A B C D E

Wrong marks should be completely erased with a clean rubber.

DO NOT FOLD THIS SHEET

1	A	B	C	D	E	21	A	B	C	D	E
2	A	B	C	D	E	22	A	B	C	D	E
3	A	B	C	D	E	23	A	B	C	D	E
4	A	B	C	D	E	24	A	B	C	D	E
5	A	B	C	D	E	25	A	B	C	D	E
6	A	B	C	D	E	26	A	B	C	D	E
7	A	B	C	D	E	27	A	B	C	D	E
8	A	B	C	D	E	28	A	B	C	D	E
9	A	B	C	D	E	29	A	B	C	D	E
10	A	B	C	D	E	30	A	B	C	D	E
11	A	B	C	D	E	31	A	B	C	D	E
12	A	B	C	D	E	32	A	B	C	D	E
13	A	B	C	D	E	33	A	B	C	D	E
14	A	B	C	D	E	34	A	B	C	D	E
15	A	B	C	D	E	35	A	B	C	D	E
16	A	B	C	D	E	36	A	B	C	D	E
17	A	B	C	D	E	37	A	B	C	D	E
18	A	B	C	D	E	38	A	B	C	D	E
19	A	B	C	D	E	39	A	B	C	D	E
20	A	B	C	D	E	40	A	B	C	D	E

文化會社出版社
投考公務員 模擬試題王

英文運用
模擬試卷（七）

時間：四十分鐘

考生須知：

（一）細讀答題紙上的指示。宣布開考後，考生須首先於適當位置貼上電腦條碼及填上各項所需資料。宣布停筆後，考生不會獲得額外時間貼上電腦條碼。

（二）試場主任宣布開卷後，考生請檢查試題冊及確定試題冊內的試題。最後會有「**全卷完**」的字眼。

（三）本試卷各題佔分相等。

（四）**本試卷全部試題均須回答**。為便於修正答案，考生宜用HB鉛筆把答案填畫在答題紙上。錯誤答案可用潔淨膠擦將筆痕徹底擦去。考生須清楚填畫答案，否則會因答案未能被辨認而失分。

（五）每題只可填畫**一個**答案。如填劃超過一個答案，該題將**不獲評分**。

（六）答案錯誤，不另扣分。

（七）未經許可，請勿打開試題冊。

CC-CRE-UE

I. Comprehension (10 questions)

Instruction:

This section aims to test candidates' ability to comprehend a written text. A prose passage of non-technical background is cited. Candidates are required to exercise skills in deciding on the gist, identifying main points, drawing inferences, distinguishing facts from opinion, interpreting figurative language, etc.

Passage:

Have you ever heard the sound of a hummingbird? They make a buzzing noise when they fly. They make this noise because they beat their wings so fast. They beat their wings up to 80 times a second. All that flapping makes a lot of noise. That's why we call them hummingbirds.

Hummingbirds fly in a unique way. They move their wings so fast that they can hover. This means that they can stay in one spot in the middle of the air, like a helicopter. Sometimes they fly or hover upside down. They are the only bird that flies backward.

Hummingbirds are small. One type called the bee hummingbird is the smallest bird in the world. Bee hummingbirds weigh less than a penny. They are just a little bit bigger than bees. I guess that's where they get their name.

Bee hummingbirds build tiny nests. They use cobwebs and bits of bark to make their homes. Their homes are only an inch around. This is big enough for their eggs though. Their eggs are smaller than peas. People have found these tiny nests on a clothespin.

Hummingbirds move fast. It takes lots of energy to move as fast as they do. This means that they need to eat a lot of food. Their favorite food is nectar, a sweet liquid inside of some flowers. They drink more than their own weight in nectar daily. They have to visit hundreds of flowers to get enough nectar to live. They can only store enough energy to survive through the night. They live on the edge.

Hummingbirds don't use their long beaks like straws. They have a tongue just like you. They use their tongues for eating. They flick their tongues in and out of their mouths while inside of flowers. They lap up nectar. flowers give them the energy that they need.

Hummingbirds help flowers too. They get pollen on their heads and bills when they feed. flowers use pollen to make seeds. Hummingbirds help pollen get from one flower to the next. This helps flowers make more seeds. More seeds means more flowers. More flowers means more food for hummingbirds. Isn't it nice how that works out?

Questions:

1. **Why are they called hummingbirds?**

   A. They are very light

   B. They sing when they fly

   C. Their wings make a humming sound

   D. Their song sounds like humming

2. **How do hummingbirds eat?**

   A. They drink nectar through their beaks like a straw.

   B. They chew up flower petals with their beaks.

   C. They use their heads and bills to eat pollen.

   D. They lap up nectar with their tongues.

3. **How do hummingbirds help flowers?**

   A. They drink nectar.

   B. They eat pollen.

   C. They bring pollen from one flower to the next.

   D. They plant seeds.

4. **According to the text, which does the bee hummingbird use to make nests?**

   A. straw

   B. concrete

   C. bark

   D. sticks

5. Which best describes the main idea of the fifth paragraph?

A. Hummingbirds move fast.

B. Hummingbirds like to eat nectar.

C. Hummingbirds use lots of energy and eat often.

D. Hummingbirds drink their own weight in nectar every day.

6. Which statement about bee hummingbirds is not true?

A. Bee hummingbird eggs are smaller than peas.

B. Bee hummingbirds weigh less than a penny.

C. Bee hummingbirds have built nests on clothespins.

D. Bee hummingbirds do not grow larger than bees.

7. What is unique about the way that hummingbirds fly?

A. They can fly faster than any other bird.

B. They can fly longer than any other bird.

C. They can fly forward and backward.

D. They can only fly for a few seconds at a time.

8. Which best defines the word hover as used in paragraph two?

A. To stay in one spot in the air

B. To clean an area thoroughly

C. An animal that has hooves

D. To move your wings very fast

9. **Why do flowers need pollen?**

A. flowers eat pollen.

B. Pollen attracts hummingbirds.

C. Hummingbirds eat pollen.

D. flowers use pollen to make seeds.

10. **Which title best describes the main idea of this text?**

A. Bee Hummingbirds: The World's Smallest Bird

B. Pollination: How Birds and flowers Work Together

C. Hummingbirds: Unique and Uniquely Helpful

D. Interesting Facts About Birds

II. Error Identification (10 questions)

Instruction:

Knowledge on use of the language is tested through identification of language errors which may be lexical, grammatical or stylistic.

Questions:

11. **The musical concert last night was a great success. The news was published on the newspaper this morning. It was well-written.**

    A. The musical concert last night

    B. a great success

    C. on the newspaper

    D. It was well-written

    E. No error

12. **The teacher let us paint with our pencils. But he did not allow us to do exercises in the same way. Instead, he wanted us to write with ink.**

    A. with our pencils

    B. in the same way

    C. Instead

    D. to write with ink

    E. No error

13. Last week, an unemployed man killed himself by jumping down from the roof of a flat. He was stupid. He must faced to realities.

A. an unemployed

B. from the roof of a flat

C. He was stupid

D. faced to realities

E. No error

14. Feeling bored after a day's work, I and my friend decided to go for a film show at the Cathay that night.

A. after a day's work

B. I and my friend

C. go for a film show

D. at the Cathay that night

E. No error

15. X: I know the word flour as well as the word flower.

Y: What is the difference in meaning of these two words?

A. I know the word flour

B. as well as the word flower

C. in meaning

D. of these two words

E. No error

16. **X: Lillian has been ill since a week. Do you know why?**

Y: She worked too hard. That's all.

A. since a week

B. Do you know why

C. worked too hard

D. That's all

E. No error

17. **X: That woman has been bedridden since her husband's death.**

Y: However, even before her husband died, she was said to be of poor health.

A. since her husband's death

B. her husband died

C. she was said

D. of poor health

E. No error

18. **X: There is something urgent for me to see to.**

Y: Will you be back at office, then?

X: Yes, after half-an-hour's time.

A. something urgent

B. to see to

C. back at office, then

D. after half-an-hour's time

E. No error

19. X: What did you do now, John?

Y: I'm doing some homework. The teacher wants to see it sometime tomorrow.

A. What did you do now

B. doing some homework

C. it

D. sometime tomorrow

E. No error

20. X: Do you know what Paul Li is, Robert?

Y: He is Assistant Manger of Eastern Co last year, but is now Manager of Far East Co.

A. Do you know what Paul Li is, Robert

B. He is

C. is now

D. Manager of Far East Co

E. No error

III. Sentence Completion (10 questions)

Instruction:

In this section, candidates are required to fill in the blanks with the best options given. The questions focus on grammatical use.

Question:

21. The candidate _____ when asked why he had left his last job; he did not want to admit that he had been ____ .

    A. demurred - promoted
    B. confided - banned
    C. dissembled - dismissed
    D. rejoiced - wrong
    E. hesitated - lauded

22. Tennyson was a well-loved poet; no other poet since has been so ____ .

    A. lionized
    B. attacked
    C. decried
    D. poetical
    E. abhorred

23. The parliamentary session degenerated into ____ with politicians hurling ____ at each other and refusing to come to order.

A. mayhem - banter

B. disarray - pleasantries

C. tranquillity - invectives

D. chaos - aphorisms

E. anarchy - insults

24. The admiral ____ his order to attack when he saw the white flag raised by the enemy sailors; he was relieved that he could bring an end to the ____ .

A. reiterated - hostilities

B. countermanded - fighting

C. commandeered - truce

D. renounced - hiatus

E. confirmed - aggression

25. In a fit of ____ she threw out the valuable statue simply because it had belonged to her ex-husband.

A. pique

B. goodwill

C. contrition

D. pedantry

E. prudence

26. **Many 17th century buildings that are still in existence have been so ____ by successive owners that the original layout is no longer ____ .**

A. preserved - visible

B. modified - apparent

C. decimated - enshrouded

D. salvaged - required

E. neglected - appropriate

27. **Since ancient times sculpture has been considered the ____ of men; women sculptors have, until recently, consistently met with ____.**

A. right - acceptance

B. domain - approbation

C. domicile - ridicule

D. realm - condolence

E. prerogative - opposition

28. **____ action at this time would be inadvisable; we have not yet accumulated sufficient expertise to warrant anything other than a ____ approach.**

A. precipitate - cautious

B. hesitant - wary

C. vacillating - circuitous

D. decisive - firm

E. ponderous - direct

29. Many biologists have attempted to ____ the conditions on earth before life evolved in order to answer questions about the ____ of biological molecules.

A. mimic - fitness

B. standardize - shapes

C. replicate - reactions

D. simulate - origin

E. ameliorate - evolution

30. Harding was unable to ____ the results of the survey; although entirely unexpected, the figures were obtained by a market research firm with an ____ reputation.

A. accept - peerless

B. discount - impeccable

C. fault - mediocre

D. counter - unenviable

E. believe - fine

IV. Paragraph Improvement (10 questions)

Instruction:

In this section, two draft passages are cited. For each passage, questions are set to test candidates' skills in improving the draft. The focus of the questions is on writing skills, not power of understanding.

Passage 1

(1)Research in education has of recent years turned its attention to the problem of assessing a student's progress as accurately as possible. (2)Only when this is done, can a meaningful course of study be laid down and a student helped to recognise his difficulties and overcome it. (3)But, despite all the new thinking on this subject, there is still no acceptable alternative to the examination. (4)There have been remarkable advances in analysing the process of learning and in framing tests and maintaining week to week records which provide the teacher with valuable information about the progress of his class and of each individual in it. (5)Whether it be at the end of term, or the year, or at the end of a school course or for the purpose of choosing candidates for a course of study and training, the only practicable way of measuring a student's performance or of assessing his potential is by an examination, supplemented where necessary, by recommendation, interview and other devices.

(6)The most unfortunate by-product of this has been the proliferation of study notes, guides to passing examination, model answers, hints for writing essays and similar travesties of education. (7)There is no need to engage on the unethical nature of these publications. (8) From the student's point of view, a rigorous censorship of this kind of publication would be a great advantage. (9)For one thing, these 'notes' promote the habit of rote learning. (10)For the other, they are priced more highly than the poems of Wordsworth or the plays of Shakespeare. (11)They are not worth the paper they are printed on.

Questions:

31. Which among the following is the best revision of the underlined portion of sentence 1 below?

Research in education has of recent years turned its attention to

A. Researchers in education has of recent years turned their attention to

B. Researchers in education have of recent years turned their attention to

C. Research in education has in recent years turned its attention to D. Research in education has of recent years turned attention to

32. Which among the following is the best revision of the underlined portion of sentence 2 below?

and a student helped to recognise his difficulties and overcome it.

A. and students helped to recognise their difficulties and overcome it.

B. and the students helped to recognise their difficulties and overcome it.

C. and a student helped to recognise his many a difficulty and overcome it.

D. and a student helped to recognise his difficulties and overcome them.

33. Which word/phrase, if inserted at the beginning of sentence 11, can help combine sentences 10 and sentence 11?

A. and yet

B. although

C. in spite of

D. despite that

34. In context, what is the best way to deal with sentence 3?

A. Move it between sentences 1 and sentence 2

B. Move it between sentences 4 and sentence 5

C. Delete it altogether.

D. Move it to the beginning of the second paragraph.

35. Which among the following best replaces the word "this" in sentence 6?

A. The student`s performance

B. The recommendation

C. The interview

D. The examination system

Passage 2

(1)Art probably owes more to form for its range of expression than to colour. (2)Many of the noblest things it is capable of conveying are expressed by form more directly than by anything else. (3)And it is interesting to notice how some of the world's greatest artists have been very restricted in their use of color, preferring to depend on form for their chief appeal. (4)It is reported that Apelles only used three colors, black, red, and yellow, and Rembrandt used little else. (5)Drawing, although the first, is also the last thing the painter usually studies. (6)There is more in it that can be taught and that requires constant application and effort. (7)A student should set himself to acquire well-trained eye of which he might be capable of; for the appreciation of every form of art. (8)It is not enough in artistic drawing to portray accurately. (9)But to express any form, one must first be moved by it. (10)There is in the appearance of all objects, animate and inanimate, a hidden rhythm that is not caught by the accurate, painstaking, but cold artist. (11)This form is never found in a mechanical reproduction like a photograph. (12)You are never moved to say when looking at one, "What fine form" . (13)It is difficult to say in what this quality consists. (14)The emphasis and selection that is unconsciously given in a drawing, done directly under the guidance of strong feeling, are too subtle to be tabulated. (15)But it is this selection of the significant and suppression of the non-essential that often gives to a few lines drawn quickly, and having a somewhat remote relation to the complex appearance of the real object, more vitality and truth than are to be found in a highly-wrought and painstaking drawing, during the process of which the essential and vital things have been lost sight of in the labour of the work; and the non-essential, which is usually more obvious, is allowed to creep in and obscure the original impression.

Questions:

36. In the context, which of the following sentences would best fit in between sentences 4 and 5?

A. Colour would seem to depend much more on a natural sense and would be less amenable to teaching.

B. Every painter should learn about both colour and drawing.

C. Colour depends more on natural sense, whereas drawing requires effort as it is difficult to understand and learn.

D. A majority of artists focus more on drawing than on colour.

37. In the context, which among the following is the best way to rephrase sentence 7(reproduced below)?

A student should set himself to acquire well-trained eye of which he might be capable of; for the appreciation of every form of art.

A. Leave it as it is now.

B. A well-trained eye for the appreciation of form is what every student should set himself to acquire with all the might of which he is capable.

C. To appreciate every form of art, all the students should acquire well-trained eyes which they might be capable of.

D. Only well-trained eyes appreciate every form of art and this is what every student requires to learn drawing.

38. What is the best way to deal with sentence 9?

A. Leave it as it is.

B. Connect it to sentence 8 by inserting "as" in place of "but".

C. Connect it to sentence 10 by inserting "since".

D. Delete it.

39. Which phrase, if inserted at the beginning of sentence 11(reproduced below), best fits the context?

This form is never found in a mechanical reproduction like a photograph.

A. Thus,

B. However,

C. In fact,

D. Also,

40. Of the following, which is the best way to revise the underlined portion of sentence 6(reproduced below)?

There is more in it that can be taught and that requires constant application and effort.

A. That can be taught and that requires constant application and effort

B. Delete "that" after "and"

C. Replace "that" after "it" with "than"

D. Replace "that" at both places with "which"

- END OF PAPER -

CC-CRE-UE

文化會社出版社 **CULTURE CROSS LIMITED**

答題紙 ANSWER SHEET

請在此貼上電腦條碼 Please stick the barcode label here

(1) 考生編號 Candidate No.

(2) 考生姓名 Name of Candidate

(3) 考生簽署 Signature of Candidate

宜用H.B.鉛筆作答
You are advised to use H.B. Pencils

考生須依照下圖所示填畫答案：

23 A B C D E

錯填答案可使用潔淨膠擦將筆痕徹底擦去。

切勿摺皺此答題紙

Mark your answer as follows:

23 A B C D E

Wrong marks should be completely erased with a clean rubber.

DO NOT FOLD THIS SHEET

No.						No.					
1	A	B	C	D	E	21	A	B	C	D	E
2	A	B	C	D	E	22	A	B	C	D	E
3	A	B	C	D	E	23	A	B	C	D	E
4	A	B	C	D	E	24	A	B	C	D	E
5	A	B	C	D	E	25	A	B	C	D	E
6	A	B	C	D	E	26	A	B	C	D	E
7	A	B	C	D	E	27	A	B	C	D	E
8	A	B	C	D	E	28	A	B	C	D	E
9	A	B	C	D	E	29	A	B	C	D	E
10	A	B	C	D	E	30	A	B	C	D	E
11	A	B	C	D	E	31	A	B	C	D	E
12	A	B	C	D	E	32	A	B	C	D	E
13	A	B	C	D	E	33	A	B	C	D	E
14	A	B	C	D	E	34	A	B	C	D	E
15	A	B	C	D	E	35	A	B	C	D	E
16	A	B	C	D	E	36	A	B	C	D	E
17	A	B	C	D	E	37	A	B	C	D	E
18	A	B	C	D	E	38	A	B	C	D	E
19	A	B	C	D	E	39	A	B	C	D	E
20	A	B	C	D	E	40	A	B	C	D	E

文化會社出版社
投考公務員 模擬試題王

英文運用
模擬試卷（八）

時間：四十分鐘

考生須知：

（一）細讀答題紙上的指示。宣布開考後，考生須首先於適當位置貼上電腦條碼及填上各項所需資料。宣布停筆後，考生不會獲得額外時間貼上電腦條碼。

（二）試場主任宣布開卷後，考生請檢查試題冊及確定試題冊內的試題。最後會有「**全卷完**」的字眼。

（三）本試卷各題佔分相等。

（四）**本試卷全部試題均須回答**。為便於修正答案，考生宜用HB鉛筆把答案填畫在答題紙上。錯誤答案可用潔淨膠擦將筆痕徹底擦去。考生須清楚填畫答案，否則會因答案未能被辨認而失分。

（五）每題只可填畫**一個**答案。如填劃超過一個答案，該題將**不獲評分**。

（六）答案錯誤，不另扣分。

（七）未經許可，請勿打開試題冊。

CC-CRE-UE

I. Comprehension (10 questions)

Instruction:

This section aims to test candidates' ability to comprehend a written text. A prose passage of non-technical background is cited. Candidates are required to exercise skills in deciding on the gist, identifying main points, drawing inferences, distinguishing facts from opinion, interpreting figurative language, etc.

Passage:

What's fiercer than a lion but smaller than a beagle? The honey badger, one of the toughest mammals in Africa and western Asia. Honey badgers stand less than a foot high. They are only a couple feet long. They weigh just over 20 pounds. Yet they have a reputation for toughness that is far greater than their size. Some honey badgers will chase away lions and take their kills. I guess that goes to show you that size isn't the only thing that matters in a fight.

So what makes the honey badger so tough? They have speed, stamina, and agility, but so do many animals. They aren't stronger than lions, so how do they stop them? The thing that sets the honey badger apart is their skin. Their skin is thick and tough. Arrows, spears, and bites from other animals can rarely pierce it. Small bullets can't even penetrate it. Not only is their skin thick and tough, it is also loose. This allows them to twist and turn to attack while another animal is gripping them. The only safe grip one can get on a honey badger is on the back of their necks.

Honey badgers have long, sharp claws. These claws are good for attacking and even better for digging. Honey badgers are some of nature's most skilled diggers. They can dig a nine-foot tunnel into hard ground in about 10 minutes. They love to catch a meal by digging up the burrows of frogs, rodents, and cobras. They also use their digging skills to create their homes. They live in small chambers in the ground and defend them fiercely. They will attack horses, cows, and even water buffalo if they are foolish enough to poke around a honey badger's den.

You don't get a reputation like the honey badger by running from danger. The honey badger is fearless and a tireless fighter. They will attack any creature that threatens them, man included. Because of the honey badger's reputation, most predators avoid them. Some animals use the honey badger's rep to their advantage. Adult cheetahs have spotted coats, but their kittens have silver manes and look like honey badgers. Some scientists believe that their coloring tricks predators into avoiding them. Wouldn't you walk the other way if you saw a honey badger?

You might be wondering: "If honey badgers are so tough, how did they get a name that makes them sound like a piece of candy?" The answer makes sense. Since honey badgers have such thick skin, bee stings rarely harm them. So honey badgers love to raid beehives. I can't blame them. Who doesn't like free honey? Honey badgers chase after honey aggressively. So much so that beekeepers in Africa have to use electric fencing to hold them back. There's nothing sweet about that.

Beekeepers aren't the only people who have grown to hate honey badgers. Honey badgers may be fun to read about, but they are nasty neighbors. They attack chickens, livestock, and some say children, though they usually leave people alone. But if a honey badger moves in your backyard, there's not a whole lot that you can do about it. I mean, are you going to go and tangle with an animal that eats the bones of its prey? An animal with teeth strong enough to crunch through turtle shells? An animal that never tires, gives up, or backs down? Yeah, I wouldn't either...

Questions:

1. Which best expresses the main idea of the third paragraph?

A. Honey badgers have sharp claws that they use for fighting.

B. Honey badgers digging skills assist them in many ways.

C. Honey badgers use their claws to defend their homes.

D. Honey badgers will defend their homes to the death against any animal.

2. **Which statement would the author most likely agree with?**

A. What makes the honey badger so tough is their speed and strength.

B. Honey badgers are large in size and tireless in fighting spirit.

C. What makes honey badgers so tough is their thick, loose skin.

D. Honey badgers got their name from the sweet taste of their meat.

3. **Which best defines the meaning of the word burrows as it is used in the third paragraph?**

A. Lily pads or other seaweeds in which animals hide

B. Holes or tunnels in which animals live

C. A nest or animal dwelling in a tree or bush

D. A water supply where small animals come to drink

4. **Which best expresses the main idea of the last paragraph?**

A. Honey badgers are a nuisance to the neighborhood.

B. Beekeepers and honey badgers do not get along well.

C. Honey badgers have very strong jaws and teeth.

D. Honey badgers eat chicken and livestock.

5. **Which best describes one of the author's main purposes in writing this text?**

   A. To persuade readers to join the efforts to protect honey badgers

   B. To compare and contrast honey badgers with beagles and lions

   C. To describe how honey badgers select their partners

   D. To explain why honey badgers are so tough

6. **Which statement would the author most likely disagree with?**

   A. Honey badgers like to raid beehives to eat honey.

   B. Honey badgers are not the biggest animals, but they may be the toughest.

   C. Honey badgers disguise their young to look like cheetah kittens.

   D. Honey badgers are not afraid to fight with humans.

7. **Which person is most likely to be disturbed by a honey badger moving in next door?**

   A. A beekeeper

   B. A biologist

   C. A bus driver

   D. A salesman

8. **Which animal is the honey badger afraid to attack?**

A. Lion

B. Water buffalo

C. Poisonous snake

D. None of these

9. **Which is not one of the honey badger's strengths?**

A. Thick skin

B. Powerful jaws and strong teeth

C. Poisonous claws

D. Tireless fighting spirit

10. **Which title best expresses the main idea of this text?**

A. Battle on the Savannah: Honey Badgers Vs. Lions

B. Little Badger, Big fight: One of Nature's Toughest Scrappers

C. Ace in the Hole: How Honey Badgers Build and Protect Their Homes

D. Little Game: Interesting Animals That Live in Africa

II. Error Identification (10 questions)

Instruction:

Knowledge on use of the language is tested through identification of language errors which may be lexical, grammatical or stylistic.

Questions:

11. **The school has organized some new language courses with a view to promote the language ability of the learners. As a result, those who took the courses are able to study more efficiently.**

    A. with a view to promote

    B. As a result

    C. who took the courses

    D. more efficiently

    E. No error

12. **I went to the United States in 1969. But I did not stay there long. I return soon after to resume duties in Singapore.**

    A. in 1969

    B. there long

    C. return

    D. resume duties in Singapore

    E. No error

13. We know full well that Andrew is going to marry with Lucy. However, nobody knows when the wedding is or where it will take place.

A. We know full well

B. marry with Lucy

C. when the wedding is

D. where it will take place

E. No error

14. The teacher asked each of us to tell the story in English with our own words, and then write it down in a nutshell.

A. each of us

B. in English

C. with our own words

D. in a nutshell

E. No error

15. X: Did you hear anything special?

Y: Yes, I heard the armed man told the landlady that if she shouted she would be shot.

A. anything special

B. the armed

C. told the landlady

D. she would be shot

E. No error

16. **X: What's your father? Where does he work every day?**

Y: My father is a hawker. He worked at a hawker centre.

A. What's your father

B. Where does he work

C. a hawker

D. He worked at a hawker centre

E. No error

17. **X: I've prepared for the test for tomorrow morning. Have you done so?**

Y: Yes, of course. I've done my work some time ago.

A. I've prepared for

B. Have you done so?

C. Yes, of course

D. I've done my work

E. No error

18. **X: My friend came while I watched TV.**

Y: You were annoyed, weren't you?

A. My friend came

B. while I watched TV

C. You were annoyed

D. weren't you

E. No error

19. **X: Some of the students talked in class when the teacher walked in.**

Y: Did the teacher scold them or pay no attention at all?

A. talked in class

B. walked in

C. scold them

D. pay no attention at all

E. No error

20. **X: What time did you go to school in this morning?**

Y: I went there at half past seven. I was in time for the first lesson.

A. go to school

B. in this morning

C. there at half past seven

D. in time for the first lesson

E. No error

III. Sentence Completion (10 questions)

Instruction:

In this section, candidates are required to fill in the blanks with the best options given. The questions focus on grammatical use.

Questions:

21. The success of the business venture ____ his expectations; he never thought that the firm would prosper.

    A. confirmed

    B. belied

    C. nullified

    D. fulfilled

    E. ratified

22. For centuries there was no ___ between their descendents; in fact ____ strife continued until modern times.

    A. peace - internecine

    B. hostility - intermittent

    C. malevolence - intense

    D. amity - contrived

    E. difference - feudal

23. The journalist ____ the efforts of the drug squad to control drug peddling, claiming that they had actually ____ the problem.

    A. commended - increased

    B. lauded - intensified

    C. decried - solved

    D. deprecated - exacerbated

    E. noted - caused

24. Since the Romans failed to ____ the tribes in Northern Britain, they built a wall to ____ the tribes.

A. conquer - alienate

B. impress - intimidate

C. subjugate - exclude

D. pacify - enrage

E. neutralize - barricade

25. The professor became increasingly ____ in later years, flying into a rage whenever he was opposed.

A. taciturn

B. voluble

C. subdued

D. contrite

E. irascible

26. Although the deep sea has a typically ____ fauna, near vents in the sea bed where warm water emerges live remarkable densities of invertebrates and fish.

A. verdant

B. unique

C. lush

D. pallid

E. sparse

27. **Their bantering talk seemed ____ , but in fact it masked an underlying ____ .**

A. hostile - antipathy

B. amicable - antagonism

C. jovial - assumptions

D. exasperating - frustrations

E. friendly - geniality

28. **The new nomenclature was so ____ that many chemists preferred to revert to the older trivial names that were at least shorter.**

A. succinct

B. cumbersome

C. irrational

D. facile

E. systematic

29. **Even though the auditors ____ the accountant, his reputation was ____ by the allegations of fraud.**

A. vindicated - enhanced

B. indicted - blemished

C. betrayed - ruined

D. exonerated - tarnished

E. cleared - condoned

30. **Many so-called social playwrights are distinctly ____ ; rather than allowing the members of the audience to form their own opinions, these writers force a viewpoint on the viewer.**

A. conciliatory

B. prolific

C. iconoclastic

D. didactic

E. contumacious

IV. Paragraph Improvement (10 questions)

Instruction:

In this section, two draft passages are cited. For each passage, questions are set to test candidates' skills in improving the draft. The focus of the questions is on writing skills, not power of understanding.

Passage 1

(1)The popular notion about marriage, that it springs from the same motives, and covers the same human needs is synonymous with love. (2)Like most popular notions this also rests not on actual facts, but on superstition.

(3)Marriage and love have nothing in common. (4)They are as far apart as the poles are; in fact, antagonistic to each other. (5)No doubt some marriages have been the result of love. (6)Not love could assert itself only in marriage; much rather it is because few people can completely outgrow a convention. (7)There are today large numbers of men and women to whom marriage is nothing but a farce, but who submit to marriage for the sake of public opinion. (8)At any rate, while it is true that some marriages are based on love, and while it is equally true that in some cases love continues in married life, I maintain that it does so regardless of marriage, and not because of it. (9)Marriage is primarily an economic arrangement, an insurance pact. (10)It differs from the ordinary life insurance agreement only in that it is more binding, more exacting. (11)Its returns are insignificantly small compared with the investments. (12)If you take out an insurance policy, dollars and cents must be paid by you. (13)If woman's premium, however, is her husband, she pays for it with her name, her privacy, her self-respect, her very life. (14)Man too pays his toll, but as his sphere is wider, marriage does not limit him as much as woman.

Questions:

31. Which of the following could be added after "not" , in sentence 6 to clarify the relationship between sentences 5 and 6?

A. however, that

B. because

C. that

D. since

32. Which sentence would be the most appropriate to follow sentence 14?

A. Marriage is thus like an insurance contract.

B. Man is relatively free and can easily move out of marriage.

C. In conclusion, only woman makes sacrifices in the relationship.

D. He feels his chains more in an economic sense.

33. In context, which of the following is the best way to phrase sentence 13(reproduced below)?

If woman's premium, however, is her husband, she pays for it with her name, her privacy, her self-respect, her very life.

A. (As it is now)

B. If, however, woman's premium is her husband, she pays for it with her name, her privacy, her self-respect, her very life.

C. If woman's premium is her husband, she pays for it with her name, privacy, self-respect, her very life.

D. If woman's premium is her husband, she has to pay for it with her name, her privacy, her self-respect, her very life.

34. In context which of the following is the best way to phrase the underlined portion of sentence 12(reproduced below)?

If you take out an insurance policy dollars and cents must be paid by you.

A. (As it is now)

B. the dollars and cents must be paid

C. you must pay the dollars and cents

D. one will have to make the payment

35. Which is the best version of sentence 1 (reproduced below)?

The popular notion about marriage, that it springs from the same motives, and covers the same human needs is synonymous with love.

A. NO CHANGE

B. The popular notion about marriage and love is that they are synonymous, that they spring from the same motives, and cover the same human needs.

C. The popular notion of love and marriage that they spring from the same motives, and cover the same human needs makes them synonymous.

D. Love and marriage spring from the same motives, and cover the same human needs and this has become a popular notion.

Passage 2

(1)These reptiles were very different from animals we are familiar with and they went extinct millions of years ago. (2)Some seem almost like monsters. (3)They existed as proved by the fossil record. (4)The young-earth fascination with dinosaurs centers on "proving" they existed with man, hence indicating their world is not ancient. (5)The next article discuses how the theory that a global flood deposited the fossil layers does not hold water. (6)People who are familiar with paleontology or history realize that before the dinosaur fossil discoveries of the 1800s, no one had ever heard of the long lost reptiles.

(7)The first evidence comes from mythology. (8)Yes, you read that right. (9)Creationists have had a history through mythology, looking for references to dragons and another beasts - and another beasts - and even the Loch Ness monster - and pointing to these things as possible indicators of dinosaurs. (10)Only recently did many creationists start agreeing with their critics that using mythology does not help one's scientific theories very will and some have stopped using these "proofs". (11)Most have also abandoned using the "Loch Ness Monster" and similar frauds and urban legends as evidences. (12)As a sign of the poor dissemination of information in creationist circles, some still print these things as true. (13)So what do young-earth creationists purport to be the evidence of dinosaurs living alongside man? (14)Even the long ago abandoned claim that human footprints were found alongside dinosaur prints in the Paluxy River bed in Texas is resurrected quite often. (15)Both secular and Christian researchers (including many young-earthers) have concluded these prints are not human, but dinosaur prints like the ones they are found with. (16)Some reasons that the "human" prints are not human: 1. Too far apart to be human: 2. Most are too large: 3. Many show claw marks, etc.; and 4. Some of the "prints" are simply erosion patterns.

Questions:

36. According to the context, which of the following is the best phrase that can be inserted at the beginning of sentence 12?

A. Yet,

B. In any case,

C. Notwithstanding

D. Not so significant

37. Which of the following is the best way to revise and combine the underlined portions of sentences 2 and 3(reproduced below)?

Some seem almost like monsters. They existed as proved by the fossil record.

A. monsters; they existed as proved by the fossil record.

B. monsters, and the fossil record proves their existence.

C. monsters as proved by the existence of fossil record.

D. monsters; yet the fossil record proves their existence.

38. Which of the following should be done with sentence 13?

A. Leave it as it

B. Insert the word, “According to you” at the beginning.

C. Delete it

D. Move lit to the end of paragraph 1 (after sentence 6)

39. Which of the following, if inserted before sentence 1, would make a good introduction to the essay?

A. People are fascinated by dinosaurs.

B. Dinosaurs and the flying reptiles were dominant during the Jurassic.

C. In the light of the lack of tangible evidence, the focus is placed on the claim that the Bible refers to dinosaurs.

D. How many asteroids did it take to wipe out the dinosaurs?

40. The best way to describe the relationship of sentence 2 to sentence 1 is that sentence 2

A. repeats the idea presented in sentence 1

B. provides explanation for sentence 1

C. corrects an inaccuracy stated in sentence 1

D. anticipates a reader's possible response to sentence 1

- END OF PAPER -

CC-CRE-UE

文化會社出版社 **CULTURE CROSS LIMITED**

答題紙 ANSWER SHEET

請在此貼上電腦條碼
Please stick the barcode label here

(1) 考生編號 Candidate No.

(2) 考生姓名 Name of Candidate

(3) 考生簽署 Signature of Candidate

宜用H.B.鉛筆作答
You are advised to use H.B. Pencils

考生須依照下圖所示填畫答案：

23 A B C D E

錯填答案可使用潔淨膠擦將筆痕徹底擦去。

切勿摺皺此答題紙

Mark your answer as follows:

23 A B C D E

Wrong marks should be completely erased with a clean rubber.

DO NOT FOLD THIS SHEET

No.	Options	No.	Options
1	A B C D E	21	A B C D E
2	A B C D E	22	A B C D E
3	A B C D E	23	A B C D E
4	A B C D E	24	A B C D E
5	A B C D E	25	A B C D E
6	A B C D E	26	A B C D E
7	A B C D E	27	A B C D E
8	A B C D E	28	A B C D E
9	A B C D E	29	A B C D E
10	A B C D E	30	A B C D E
11	A B C D E	31	A B C D E
12	A B C D E	32	A B C D E
13	A B C D E	33	A B C D E
14	A B C D E	34	A B C D E
15	A B C D E	35	A B C D E
16	A B C D E	36	A B C D E
17	A B C D E	37	A B C D E
18	A B C D E	38	A B C D E
19	A B C D E	39	A B C D E
20	A B C D E	40	A B C D E

文化會社出版社
投考公務員 模擬試題王

英文運用
模擬試卷(九)

時間：四十分鐘

考生須知：

(一) 細讀答題紙上的指示。宣布開考後，考生須首先於適當位置貼上電腦條碼及填上各項所需資料。宣布停筆後，考生不會獲得額外時間貼上電腦條碼。

(二) 試場主任宣布開卷後，考生請檢查試題冊及確定試題冊內的試題。最後會有「**全卷完**」的字眼。

(三) 本試卷各題佔分相等。

(四) **本試卷全部試題均須回答**。為便於修正答案，考生宜用HB鉛筆把答案填畫在答題紙上。錯誤答案可用潔淨膠擦將筆痕徹底擦去。考生須清楚填畫答案，否則會因答案未能被辨認而失分。

(五) 每題只可填畫**一個**答案。如填劃超過一個答案，該題將**不獲評分**。

(六) 答案錯誤，不另扣分。

(七) 未經許可，請勿打開試題冊。

CC-CRE-UE

I. Comprehension (10 questions)

Instruction:

This section aims to test candidates' ability to comprehend a written text. A prose passage of non-technical background is cited. Candidates are required to exercise skills in deciding on the gist, identifying main points, drawing inferences, distinguishing facts from opinion, interpreting figurative language, etc.

Passage:

You know that you're doing something big when your company name becomes a verb. Ask Xerox. In 1959 they created the first plain paper copy machine. It was one of the most successful products ever. The company name Xerox grew into a verb that means "to copy", as in "Bob, can you Xerox this for me?" Around 50 years later, the same thing happened to Google. Their company name grew into a verb that means "to do an internet search". Now everyone and their grandma knows what it means to Google it.

Unlike Xerox, Google wasn't the first company to invent their product, not by a long shot. Lycos released their search engine in 1993. Yahoo! came out in 1994. AltaVista began serving results in 1995. Google did not come out until years later, in 1998. Though a few years difference may not seem like much, this is a major head start in the fast moving world of tech. So how did Google do it? How did they overtake their competitors who had such huge leads in time and money? Maybe one good idea made all the difference.

There are millions and millions of sites on the internet. How does a search engine know which ones are relevant to your search? This is a question that great minds have been working on for decades. To understand how Google changed the game, you need to know how search engines worked in 1998. Back then most websites looked at the words in your query. They counted how many times those words appeared on each page. Then they might return pages where the words in your query appeared the most. This system did not work well and people often had to click through pages and pages of results to find what they wanted.

Google was the first search engine that began considering links. Links are those blue underlined words that take you to other pages when you click on them. Larry Page, cofounder of Google, believed that meaningful data could be drawn from how those links connect. Page figured that websites with many links pointing at them were more important than those that had few. He was right. Google's search results were much better than their rivals. They would soon become the world's most used search engine.

It wasn't just the great search results that led to Google becoming so well liked. It also had to do with the way that they presented their product. Most of the other search engines were cluttered. Their home pages were filled with everything from news stories to stock quotes.

But Google's homepage was, and still is, clean. There's nothing on it but the logo, the search box, and a few links. It almost appears empty. In fact, when they were first testing it, users would wait at the home page and not do anything. When asked why, they said that they were, "waiting for the rest of the page to load" . People couldn't imagine such a clean and open page as being complete. But the fresh design grew on people once they got used to it.

These days Google has its hands in everything from self-driving cars to helping humans live longer. Though they have many other popular products, they will always be best known for their search engine. The Google search engine has changed our lives and our language. Not only is it a fantastic product, it is a standing example that one good idea (and a lot of hard work) can change the world.

Questions:

1. **Which event happened last?**

   A. Lycos released their search engine.

   B. Yahoo! released their search engine.

   C. Google released their search engine.

   D. Xerox released their copy machine.

2. **Which statement would the author of this text most likely disagree with?**

   A. Part of Google's success is due to the design of their homepage.

   B. Google succeeded by following examples of others in their field.

   C. Google wasn't the first search engine, but it was the best.

   D. Google's success may not have been possible without Larry Page.

3. **Which best expresses the main idea of the third paragraph?**

   A. There are lots and lots of websites connected to the internet.

   B. Google created a better way to organize search results.

   C. Many smart people have worked on search engines over the years.

   D. Older search engines used unreliable methods to order results.

4. **What is the author's main purpose in writing this article?**

   A. To explain how Google overtook its rivals

   B. To compare and contrast Google and Xerox

   C. To persuade readers to use Google for internet searches

   D. To discuss how companies can influence language over time

5. **Which statement would the author most likely agree with?**

   A. Google became successful because its founders were well-connected.

   B. Google was the world's first and best search engine.

   C. Google changed the world by solving an old problem in a new way.

   D. Google's other products are now more important to its success than search.

6. **Which best expresses the main idea of the fourth paragraph?**

   A. Links allow people to surf from one website to the next.

   B. Larry Page's ideas about links helped Google get to the top.

   C. Larry Page contributed to the internet by inventing the link.

   D. Google is a website that serves important links to users.

7. **Which best explains why the author discusses Xerox in this text?**

   A. He is discussing big companies that came before Google.

   B. He is explaining how companies must change with the times.

   C. He is showing how companies can affect our language.

   D. He is comparing and contrasting Google and Xerox.

8. **How did Google improve search quality in 1998?**

   A. They counted how many times queries appeared on each page.

   B. They looked more closely at the words in search queries.

   C. They linked to more pages.

   D. They studied the relationships of links.

9. **Which was cited as a reason why Google became so popular?**

   A. Google's homepage was clean.

   B. Google provided catchy news stories on their homepage.

   C. Google homepage loaded quickly.

   D. Google provided useful stock quotes on their homepage.

10. **Which title best expresses the author's main purpose in writing this text?**

    A. Xerox Vs. Google: Battle of the Titans

    B. Search Engines: How They Work and Why They're Important

    C. A Better Way: How Google Rose to the Top

    D. Search Engines: A Short History of Important Tools.

II. Error Identification (10 questions)

Instruction:

Knowledge on use of the language is tested through identification of language errors which may be lexical, grammatical or stylistic.

Questions:

11. **Although the answer of this question is quite easy, not many candidates can answer it to the satisfaction of the examiner.**

    A. answer of this question

    B. not many candidates

    C. it

    D. to the satisfaction of

    E. No error

12. **Some worked hard. As a consequence, they made great progresses in their studies and in their work.**

    A. Some

    B. As a consequence

    C. made great progresses

    D. in their work

    E. No error

13. **Those who work without a long-term plan is bound to fail. However, it does not mean that all such persons are doomed to failure.**

A. without a long-term plan

B. is bound to fail

C. all such persons

D. doomed to failure

E. No error

14. **Thank you for your letter of 9 August. Enclosed is our latest information for your perusal. I look forward to hear from you soon.**

A. your letter of 9 August

B. Enclosed

C. for your perusal

D. hear from you soon

E. No error

15. **X: Alan hurted himself when he fell down from the tree.**

Y: Too bad! Why was he so careless?

A. Alan hurted himself

B. from the tree

C. Too bad

D. so careless

E. No error

16. **X: My hen laid its first egg this morning.**

Y: My hens lay eggs every day. But they do not lay any this morning.

A. laid its first egg

B. hens lay eggs

C. do not lay

D. any

E. No error

17. **X: What's wrong with Mary?**

Y: Nothing. I told her if she tried harder, she would have passed.

A. wrong with Mary

B. Nothing

C. I told her

D. she tried harder

E. No error

18. **X: Listen! I'm going to tell you a story.**

Y: Can you tell us the life of a hero in ancient China?

A. Listen

B. tell you a story

C. tell us the life

D. in ancient China

E. No error

19. **Hurrying across the playground, her books fell in the mud. She then picked them up, but they were muddy.**

A. Hurrying across the playground

B. fell in the mud

C. picked them up

D. muddy

E. No error

20. **Although there is an English test tomorrow, Tom wants to go to the pictures this evening. Knowing this, his mother said. "You'd better not to go but stay at home and study."**

A. to the pictures

B. Knowing this

C. You'd better not to go

D. study

E. No error

III. Sentence Completion (10 questions)

Instruction:

In this section, candidates are required to fill in the blanks with the best options given. The questions focus on grammatical use.

Questions:

21. His one vice was gluttony and so it is not surprising that as he aged he became increasingly ____ .

    A. emaciated

    B. despondent

    C. corpulent

    D. carping

    E. lithe

22. Our once thriving High School Nature Club is now ____ ; the programs have had to be cancelled due to lack of support.

    A. defunct

    B. extant

    C. resurgent

    D. burgeoning

    E. renovated

23. Having been chief accountant for so many years, Ms. George felt herself to be ____ and was unwilling to ____ control of the department after the merger.

A. slighted - truncate

B. irreplaceable - assume

C. insubordinate - retain

D. decisive - continue

E. indispensable - relinquish

24. Because Elaine's father was a field entomologist who trekked over the continent studying insect infestations, and insisted on taking his young family with him, Elaine and her brother had a(n) ____ childhood.

A. idyllic

B. itinerant

C. sedentary

D. propitious

E. equable

25. Frederica was ____ when her supervisor took only a ___ look at her essay over which she had taken so much care.

A. exultant - superficial

B. vexed - studious

C. disappointed - cursory

D. pleased - patronizing

E. relieved - perfunctory

26. When he was young he ____ ideas of becoming a doctor; however, he was ____ by his father who wanted him to join the family business.

A. harbored - backed

B. entertained - dissuaded

C. produced - critical

D. repudiated - deterred

E. eschewed - encouraged

27. Literary criticism has in recent years become increasingly ____ ; it is almost impossible for the non-literary person to understand its analyses.

A. abstruse

B. accessible

C. colloquial

D. wide-ranging

E. professional

28. The alchemists, though they are often supposed to have been ____ or confidence tricksters, were actually skillful technologists.

A. empiricists

B. polemicists

C. pragmatists

D. theorists

E. charlatans

29. **Bullock carts and hand pumps seem ____ in a village whose skyline is dominated by telephone cables and satellite dishes.**

A. anachronisms

B. exigencies

C. diversions

D. provocations

E. portents

30. **A ____ child, she was soon bored in class; she already knew more mathematics than her junior school teachers.**

A. obdurate

B. querulous

C. precocious

D. recalcitrant

E. contemporary

IV. Paragraph Improvement (10 questions)

Instruction:

In this section, two draft passages are cited. For each passage, questions are set to test candidates' skills in improving the draft. The focus of the questions is on writing skills, not power of understanding.

Passage 1

(1)Life is a white water river. (2)At first I paddle slowly along the uncharacteristically-calm waters to school. (3)I have got to steel myself to keep up with the tiresome work that lies ahead of me.

(4)The heavy waves of work pound me tirelessly as I attempt to manoeuvre around the feared whirlpool of depression. (5)It, with time, pulls me under and swallow me alive.

(6)The river, I may call life, would be in control at all times; I must quickly steer and paddle away from the massive rocks and devouring waves. (7)When travelling along a rough river, it is essential that the rafter be a quick thinker, and only the experienced rafter can outwit the many perils that lie ahead.

(8)Sure, I am apt to make mistakes, I can only imagine how many times waves have knocked my raft over while learning, I can only imagine how many rocks my raft has crashed into when I did not receive a top mark: I can only imagine how many dreadful times my usually sturdy raft has almost been pulled into the whirlpool of depression.

(9)On the other hand, there are many positive aspects to this white water river with its exhilarating highs and its anticipated lows. (10) There are many interested people whom I may accidentally bump my raft along the way. (11)Of course, the river giving me the ultimate thrill of knowing that I have managed to succeed the dastardly rough-waters. (12)Then, as the water begins to flow smoothly again, I can relax and cherish the experience, and anticipate the dangers and hard work that lie ahead.

Questions:

31. The primary effect of the final paragraph (sentences 9 to 12) is to:

A. summarize the ideas introduced in the preceding paragraph.

B. use persuasion to change the reader's opinion.

C. continue the essay's tone of drawing an analogy.

D. explain contradictions within the essay

32. There are many interested people whom I may accidently bump my raft along the way. Of course the river giving me the ultimate thrill.... What is the best way to combine these two sentences?

A. bump my raft along the way and then the river giving me

B. bump raft when on my way, thereby river giving me

C. bump my raft into on my way and of course, the river can give me

D. bump my raft as on my way and there the river must give me

33. In context, which of the following revisions is necessary in sentence 8(reproduced below)?

Sure, I am apt to make mistakes, I can only imagine how many times waves have knocked my raft over while learning, I can only imagine how many rocks my raft has crashed into when I did not receive a top mark: I can only imagine how many dreadful times my usually sturdy raft has almost been pulled into the whirlpool of depression.

A. Leave it as it is

B. Delete “sure,”

C. Change “colon” to “comma” .

D. Replace “comma” and “colon” with “semi-colon” .

34. The best way to describe the relationship of sentence 2 to sentence 1 is that sentence 2:

A. corrects an inaccuracy stated in sentence 1.

B. elucidates the metaphor referred to in sentence 1.

C. anticipates a reader’ s possible response to sentence 1.

D. introduces a contrasting view of sentence 1.

35. In context, which is the best version of the underlined portions of sentences 4 and 5(reproduced below)?

The heavy waves of work, pound me tirelessly as I attempt to manoeuvre around the feared whirlpool of depression. It, with time, pulls me under and swallow me alive.

A. whirlpool of depression; with time, pulling

B. whirlpool of depression, which at times is pulling

C. whirlpool of depression, which at any given time can pull

D. whirlpool of depression that at number of times pull

Passage 2

(1)Blowing bubbles in my open water class, I've logged over 100 dives. (2)This love for diving has evolved into an intense passion towards protecting the ocean, and all of its inhabitants. (3)I've chosen to put my love for the ocean into action, as an environmentalist. (4) Actually, this passion extends out towards efforts that took to help all the planetary domains gain protection. (5)As such, I appreciate when others take the time educate me on those other realms for which I know less about. (6)To be an environmentalist, one must choose the cause which resonates within ones sole, and run with it. (7)One must be willing to educate people about the environment. (8)And educate those people who support other causes. (9)Together we can help each other towards learning how to become a true "Environmentalist". (10)We must all encourage positive collaboration and education as opposed to being against something. (11)For example, sharks are being decimated to near extinction simply for their fins. (12)The fins are used to make Shark fin soup, a delicacy popular particularly in Taiwan and Singapore. (13)It would be easy to blame these communities for creating the demand. (14)However in conversing with Asian environmentalists, they liken the culture around eating Shark fin soup to the culture surrounding Americans eating turkey for Thanksgiving dinner. (15)There are continuous and progressing efforts to educate these people, by members of their own community, on just how dangerous his cultural practice is and the devastating impact this could have on their (our) world if all the sharks were to disappears as a result.

(16)Famous restaurants have taken endangered Swordfish off their menus, these same restaurants are buying wild-caught salmon (and boosting the economy of local fisheries in the process), laundromats have started selling green detergent, this just to name a few of these enlightened changes. (17)This is how the "Environmentalist" can begin the revolution. (18)Just find something you believe in and make a stand. (19)One by one, we can make our planet a cleaner place to live, steeped in healthy bio-diversity for generations to come.

Questions:

36. Of the following, which is the best way to revise and combine the underlined portions of sentences 7 and 8(reproduced below)?

One must be willing to educate people about the environment. And educate those people who support other causes.

A. environment; and educate those people who support other causes.

B. environment educating those people who are supporting other causes.

C. environment besides educating those people who are in support of other causes.

D. environment while being open to education from those people who support other causes.

37. Sentence 4 in the passage is best described as:

A. summing up the theme of the passage

B. providing an additional example

C. introducing a new topic

D. presenting a personal opinion

38. In context, which is the best version of the underlined portion of sentence 1 (reproduced below)?

Blowing bubbles in my open water class, I've logged over 100 dives.

A. Leave it as it is

B. I used to blow bubbles

C. I have been blowing bubbles ever since

D. Since the first time having blown bubbles

39. Which of the following is the best sentence to insert at the beginning of the second paragraph?

A. Environmentalists everywhere are making a difference.

B. Besides my love for diving, savoring sea food is my passion.

C. Environmentalists should also obtain education on seafood cuisines.

D. To preserve the endangered species, we must switch to other options available in the market.

40. In context, which of the following revisions would NOT improve sentence 15(reproduced below)?

There are continuous and progressing efforts to educate these people, by members of their own community, on just how dangerous his cultural practice is and the devastating impact this could have on their (our) world if all the sharks were to disappears as a result.

A. Insert "In the view" at the beginning of the sentence.

B. Change "Continuous and progressing" to "ongoing"

C. Change "their own community" to "the different community"

D. Delete "this cultural practice is and the"

- END OF PAPER -

PART TWO
模擬試題答案

PAPER 1

1. A
2. B
3. D
4. C
5. C
6. B
7. B
8. B
9. D
10. A
11. B
12. B
13. D
14. D
15. B
16. B
17. B
18. B
19. B
20. B
21. D
22. E
23. D
24. C
25. D
26. A
27. B
28. E
29. B
30. C
31. C
32. A
33. D
34. C
35. B
36. B
37. D
38. A
39. C
40. A

PAPER 2

1. B
2. A
3. D
4. B
5. D
6. C
7. A
8. C
9. D
10. A
11. D
12. C
13. B
14. B
15. A
16. D
17. C
18. D
19. C
20. D
21. B
22. A
23. B
24. D
25. A
26. B
27. B
28. E
29. E
30. D
31. D
32. B
33. C
34. A
35. C
36. B
37. D
38. C
39. A
40. B

PAPER 3

1. C
2. A
3. A
4. D
5. B
6. D
7. B
8. C
9. B
10. D

11. C
12. D
13. D
14. C
15. A
16. A
17. B
18. D
19. C
20. C
21. D
22. B
23. E
24. D
25. A
26. C
27. A
28. D
29. D
30. E
31. C
32. D
33. A
34. A
35. B
36. C
37. D
38. B
39. B
40. C

PAPER 4

1. C
2. A
3. C
4. D
5. C
6. B
7. D
8. A
9. B
10. D
11. A
12. A
13. A
14. D
15. D
16. B
17. C
18. C
19. D
20. C
21. A
22. C
23. E
24. C
25. D
26. B
27. E
28. A
29. C
30. B
31. D
32. B
33. C
34. C
35. B
36. D
37. C
38. B
39. D
40. B

PAPER 5

1. D
2. B
3. A
4. C
5. C
6. D
7. A
8. B
9. D
10. B
11. D
12. C
13. D
14. B
15. D
16. C
17. C
18. D
19. C
20. B

21. A
22. C
23. D
24. B
25. C
26. B
27. B
28. D
29. C
30. A
31. B
32. B
33. D
34. D
35. B
36. B
37. C
38. B
39. A
40. A

PAPER 6

1. B
2. D
3. C
4. D
5. A
6. B
7. C
8. C
9. D
10. A
11. A
12. B
13. B
14. D
15. A
16. D
17. A
18. A
19. A
20. A
21. B
22. D
23. E
24. A
25. A
26. D
27. B
28. C
29. A
30. E
31. C
32. D
33. A
34. C
35. A
36. D
37. B
38. A
39. B
40. C

PAPER 7

1. C
2. D
3. C
4. C
5. C
6. D
7. C
8. A
9. D
10. C
11. C
12. D
13. D
14. B
15. D
16. A
17. D
18. D
19. A
20. B
21. C
22. A
23. E
24. B
25. A
26. B
27. E
28. A
29. D
30. B

31. C
32. D
33. B
34. B
35. D
36. C
37. B
38. B
39. D
40. C

PAPER 8

1. B
2. C
3. B
4. A
5. D
6. C
7. A
8. D
9. C
10. B
11. A
12. C
13. B
14. C
15. C
16. D
17. D
18. B
19. A
20. B
21. B
22. A
23. D
24. C
25. E
26. E
27. B
28. B
29. D
30. D
31. B
32. D
33. B
34. C
35. B
36. A
37. D
38. D
39. A
40. D

PAPER 9

1. C
2. B
3. D
4. A
5. C
6. B
7. C
8. D
9. A
10. C
11. A
12. C
13. B
14. D
15. A
16. C
17. D
18. C
19. A
20. C
21. C
22. A
23. E
24. B
25. C
26. B
27. A
28. E
29. A
30. C
31. B
32. C
33. D
34. B
35. C
36. D
37. A
38. D
39. D
40. B

看得喜 放不低

創出喜閱新思維

書名 《投考公務員 英文運用模擬試卷精讀》（第二版）
ISBN 978-988-76628-1-5
定價 HK$128
出版日期 2022 年 11 月
作者 Fong Sir
責任編輯 文化會社公務員系列編輯部
版面設計 Rocksteddy
出版 文化會社有限公司
電郵 editor@culturecross.com
網址 www.culturecross.com
發行 聯合新零售（香港）有限公司
地址：香港鰂魚涌英皇道 1065 號東達中心 1304-06 室
電話：（852）2963 5300
傳真：（852）2565 0919